AF477457

The Monsoon Regime of the Currents in the Indian Ocean

THE MONSOON REGIME OF THE CURRENTS IN THE INDIAN OCEAN

by Walter Düing

EAST-WEST CENTER PRESS   Honolulu

To Professor Günter Dietrich
for his inspiring, scholarly guidance and personal kindness

Contribution No. 330
from the Hawaii Institute of Geophysics,
University of Hawaii.

Acknowledgment

The author is especially indebted to K. Wyrtki, editor of the *International Indian Ocean Expedition, Oceanographic Atlas*. His comments, valuable suggestions, and constructive criticism are sincerely appreciated. Many discussions with B. Gallagher, G. Groves, and E. D. Stroup were of great help in clarifying the concepts concerning the monsoonal circulation of the Indian Ocean. These discussions also contributed largely toward making my two-year period at the University of Hawaii extremely pleasant and a most memorable event.

My thanks are also attributed to the Statistical and Computing Center of the University of Hawaii, which provided computing time without charge, and to Mr. L. Popwell, who wrote the programs and plotting routines. In addition, my deep appreciation is expressed to Mrs. C. Stroup for the invaluable help and considerable concern she provided in the editing of this manuscript.

Appreciation also is expressed to H. Taylor and R. Rhodes for their skillful assistance in preparing the illustrations and to Miss Betsy Murashige for her untiring assistance in typing the text.

The investigation was partly sponsored by the National Science Foundation under Grant No. GA–1279.

The printing of this monograph was financed by the Office of Naval Research under Contract Nonr 3748 (06), a support which is gratefully acknowledged.

Walter Düing, University of Miami

Honolulu, September 21, 1969

Contents

Abstract

All hydrographic data collected during the International Indian Ocean Expedition from 1960 to 1965, and additional data from previous years, have been compiled at the University of Hawaii so that an atlas of the physical oceanography of the Indian Ocean can be prepared. The present investigation, which is based on these data, is limited to a discussion of the effects of the monsoonal winds on the surface circulation north of 20° S. Although the accumulated data is the most comprehensive material on the Indian Ocean, it is still very heterogeneous; hence, it was necessary to apply statistical criteria in order to remove errors and variations that were introduced partly by nonsynoptic observations.

The first part of this investigation presents the dynamic topographies of the sea surface for spring, early summer, late summer, fall, and winter. In order to cover each period, it was necessary to compile average charts of the data accumulated on cruises made in different years. In addition, one chart for the Arabian Sea has been contoured using data from the summer of 1963 only. The main feature, common to all charts, is the occurrence of a complex pattern of lows and highs. A comparison with the sea-surface topographies of the Atlantic Ocean and the Pacific Ocean reveals that large gyres existing in these two oceans are not found in the northern part of the Indian Ocean and, vice versa, that the conditions prevailing in the Indian Ocean are found nowhere else. There is evidence in the average charts that the number of cyclonic or anticyclonic vortices is higher during the transition seasons than during the summer and winter season. A brief discussion of the dynamic topographies in terms of the conventionally known surface currents is given.

The second part presents a theoretical model in an attempt to interpret the peculiarities of the monsoonal circulation. The nonstationary, linear, barotropic model includes dissipation in the simple form, $k\nabla^2\psi$, and it takes into account the time-dependent components of the zonal and the meridional wind stress. Because of the linearity of the model, it is possible to solve the zonal part of the problem separately from the meridional part. A detailed discussion of the zonal solution leads to the following conclusions: The rotation of the earth acts as an externally imposed ordering mechanism causing a decreasing number of circulation cells with increasing values of β. This effect can be recognized only when very small values of the frictional factor are used. No westward intensification is produced under the absence of friction. East-west asymmetry and a decreasing number of circulation cells are achieved by increasing viscous effects. The zonal solution is interpreted as two superimposed forced Rossby waves, which have the same frequency as the wind stress. It is demonstrated that the model simulates the observations satisfactorily for the case of an inertial regime; that is, if the frictional parameter, k, is less than ω, where $\omega = 2\pi$ (year)$^{-1}$.

The zonal solution yields maximum volume transports of 50×10^6 m^3 sec^{-1} during July. A maximum phase lag of 10 days occurs between wind and volume transport in the western boundary region. The meridional solution yields comparatively small maximum-transport rates of 3×10^6 m^3 sec^{-1} and a phase lag of 70 days. Therefore, its contribution to the total transport during winter and summer is negligible in comparison with the contribution of the zonal solution. During the transition periods, however, the meridional part has considerable influence since its transport rates are then of the same order as those of the zonal part. The meridional part, furthermore, causes the circulation pattern to become very complicated during spring and fall. The most outstanding features are the rapid change in the circulation pattern and the rapid variation of the volume transport over short time intervals during the transition seasons.

The author is convinced that the average charts for the summer and winter seasons are fairly representative. He is also convinced, however, that the present methods of oceanographic survey are insufficient to deal with the extremely variable conditions during spring and fall.

Introduction

Since classical times it has been recognized that the winds over the northern Indian Ocean reverse semiannually. It is a well-known fact that the monsoons were used to great advantage by the Greek seafarers in the first centuries A.D. in carrying out their extensive Arabian Sea trade with India. From Arabic documents of the medieval period, however, we know that the semiannual reversal of the surface currents was discovered only in the ninth or tenth century. In a very readable study of early Arab references, Warren (1966) gives many interesting details about this topic.

The fact that the oldest trade routes lead through the northern Indian Ocean contributed to an accumulated knowledge of the general surface current system. After the invention of the chronometer, the seafaring nations utilized ship's drift to draw up atlases giving monthly charts of the surface currents. Figures 1 and 2 show the wind and surface current distribution during the Northeast Monsoon (northern winter) and the Southwest Monsoon (northern summer).

The periodical reversal of the wind and of the surface circulation over such an extended area is outstanding when compared to that of the Atlantic Ocean or the Pacific Ocean. The large-scale circulation in the northern part of the Indian Ocean (north of 20° S) has an essentially nonstationary character, whereas the large-scale circulation in the great Atlantic and Pacific gyres shows an essentially stationary behavior. The uniqueness of this phenomenon provides the oceanographer with a superb natural laboratory for the investigation of time-dependent processes.

But, owing to the relatively fast changes of all the hydrographic parameters and to our limited capabilities of doing a synoptic or even quasi-synoptic survey, the Indian Ocean presents an enormous problem to the organization of oceanographic observations. Only a great international effort such as the International Indian Ocean Expedition (IIOE) from 1959 to 1965 was able to accumulate sufficient data to permit a detailed treatment of the complex problems involved.

All available data on the Indian Ocean have been brought together at the University of Hawaii under the sponsorship of the National Science Foundation so that an atlas of the physical oceanography of the Indian Ocean can be prepared. At the time of writing, 10,037 hydrographic stations have been recorded. Of these, only 3,060 could be used in this report partly because many stations did not reach deep enough and partly because some had to be discarded to remove heterogeneities. This monograph is limited to a discussion of the effects of the monsoons on the surface circulation of the Indian Ocean north of 20° S. The reader who is interested in a more complete descriptive picture of the physical conditions of the entire Indian Ocean is referred to the aforementioned atlas, which is being prepared under the direction of Professor Klaus Wyrtki at the University of Hawaii.

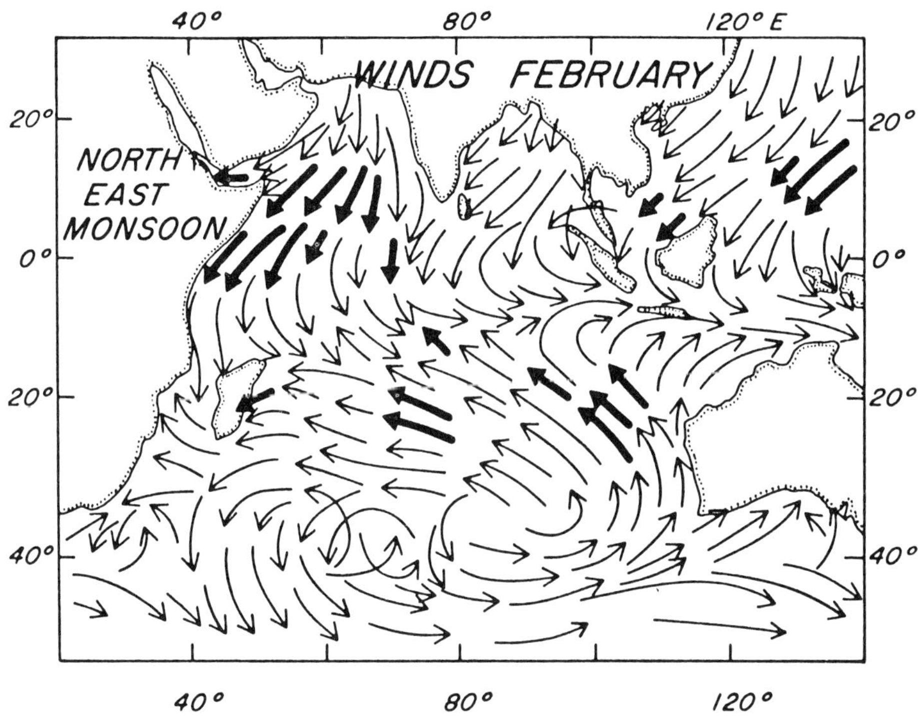

FIGURE 1: Idealized wind distributions during February and August. According to *Atlas of Climatic Charts of the Oceans* (1938).

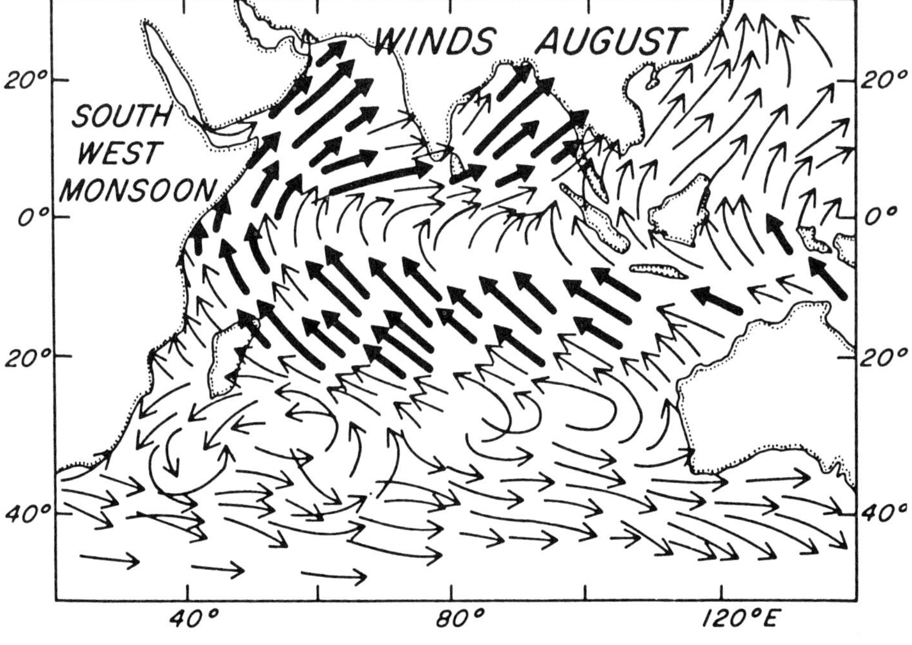

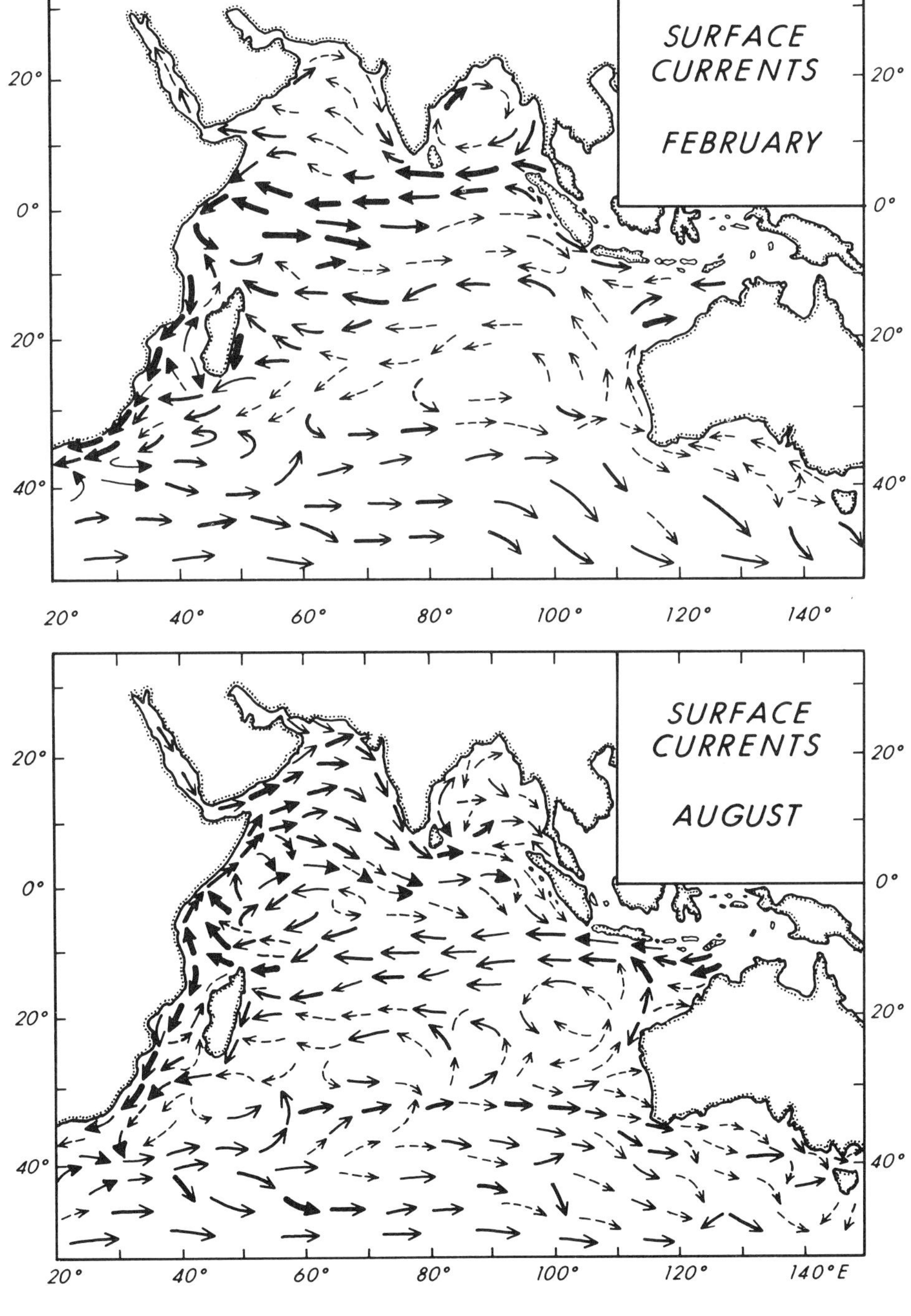

FIGURE 2:   Idealized surface current distributions during February and August. According to *Monatskarten für den Indischen Ozean*, Deutsches Hydrographisches Institut (1960).

PART 1: *Observational Results from the International Indian Ocean Expedition*

I. DEFINITION OF THE MONSOON AREA

In order to determine the extent of the region to be considered, the area of the monsoon regime will be defined in terms of atmospheric as well as oceanic parameters, in the following three ways: (i) the horizontal extent of the monsoon winds as well as the area of the monsoonally affected surface currents; (ii) the depth of influence of the monsoonal circulation; and (iii) the characteristic time periods during which the observations indicate consistent conditions. For this purpose, observations from the International Indian Ocean Expedition (1960–1965) as well as from publications dealing with the subject have been used.

A. The Horizontal Extent of the Monsoon Area

The winds. One way of defining the monsoon regime over the Indian Ocean is to consider as monsoonal those overwater areas where the direction of the average winds changes by more than 90 degrees between winter and summer. We follow the procedure of Ramage (1968), who has excluded the regions of very light winds or calms, the doldrums, where directions can fluctuate widely over short intervals. The distribution of the changes of the wind direction (Fig. 3) shows that the monsoons are essentially a phenomenon of the northern hemisphere. The diagonally hatched area in Figure 3 indicates a change of more than 90 degrees in the mean resultant wind direction between January and July. Hence, the monsoons are limited to the area north of 10° S. This is not surprising, because the southern hemisphere is also known as the *water* hemisphere, where monsoonal effects owing to the absence of meridional land barriers do not occur. Only in a few localities off the east coast of Africa and around northern Australia is the overwhelming, modifying influence of the ocean on the seasons overcome by the ocean-continent interaction that produces the monsoons.

The surface currents. In following the definition of the monsoon regime of the wind, we consider as monsoonal those areas of surface currents where the directions of the average surface currents change by more than 90 degrees between winter and summer. This definition, however, does not represent as clear cut a case for the surface currents as it does for the atmospheric conditions for two reasons: First, surface currents are only partly wind driven. In the northern part of the Indian Ocean there exists a sharp climatic contrast, the western part having an excessively dry climate, and the eastern part, an essentially humid climate. This contrast results in considerable differences in the surface salinities and temperatures between these areas, thus affecting the thermohaline circulation that is superimposed on the wind-driven circulation. Second, our knowledge of surface currents is based mainly on calculations of ship's drift where the amount of drift owing to winds can only be estimated. The diagram that defines the area on monsoonal influence (Fig. 4) is based on 5-degree averages for the surface currents during January and during July as presented in *Monatskarten für den Indischen Ozean,* Deutsches Hydrographisches Institut (1960). Keeping in mind the inherent shortcomings of these methods, we find Figures 3 and 4 agreeing surprisingly well. Over the greatest part of the Indian Ocean, the limit between the monsoonal and the *normal* circulation is drawn by the latitude circle of 10° S. Minor deviations from Figure 3 show up near the island of Madagascar and near the northwest coast of Australia. Both regions coincide with the southern limits of the monsoon wind area. Near Africa, topographic effects become important.

B. The Depth of Influence of the Monsoon Circulation

It is a far more difficult task to determine the vertical limits — than the horizontal — of the monsoonal influence. Owing to the lack of direct current measurements, such an investigation must be based on considerations of the dynamic heights as computed from temperature and salinity observations. Details of the computation of the dynamic heights, such as the number of stations used, the quality of data, and the choice of the reference level, are given in Part 1, Section II. It is sufficient to say here that the 1000 dbar level has been accepted as a reference level for the area north of 20° S.

The following procedure was used in order to gain an insight into the vertical extension of the monsoonal effects:

First, the annual variation of the dynamic heights, ΔH, was determined by computing the mean values of the dynamic heights for each 10-degree square north of 20° S. The computations were carried out for every month for the levels 0, 100, 150, 200, 300, 400, and 500 dbars. When enough data were available, the means were computed for 5-degree squares also. Although averaging over even smaller areas would have been desirable in the western boundary region of the ocean, this was not feasible, owing to the insufficient number of observations. The annual variation, ΔH, was then considered to be represented by the difference between the maximum mean value and the minimum mean value.

Three sets of typical curves, which show the vertical distribution of ΔH, for the western, central, and eastern regions, have been selected and are given in Figure 5. The values in the western region generally exceed those for the rest of the ocean. The lowest surface values are found in the central region. The curves for the eastern region are marked by a sharp drop at a depth between 100 m and 150 m.

Second, the depth of penetration of ΔH was determined. For this purpose, $\Delta H = 7.5$ dyn cm was selected as a lower limit for a notice-

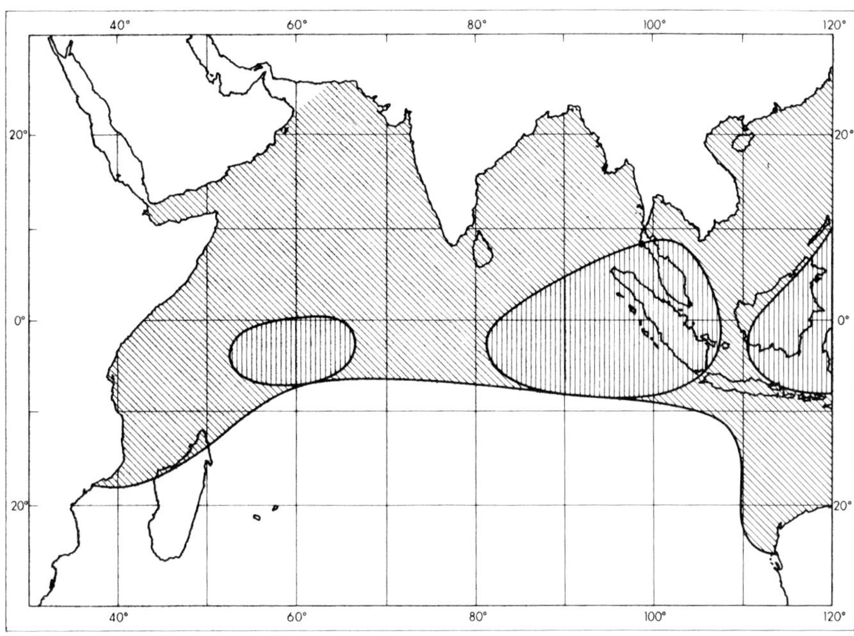

FIGURE 3: Wind direction as a monsoon criterion. A change in the mean resultant wind direction (as determined from ship reports) of more than 90° indicates that the ocean area possesses a monsoonal climate, after Ramage (1968). The diagonally hatched area indicates a change of more than 90° in mean resultant wind directions between January and July. The vertically hatched area indicates mean resultant winds of less than 5 knots during January and July.

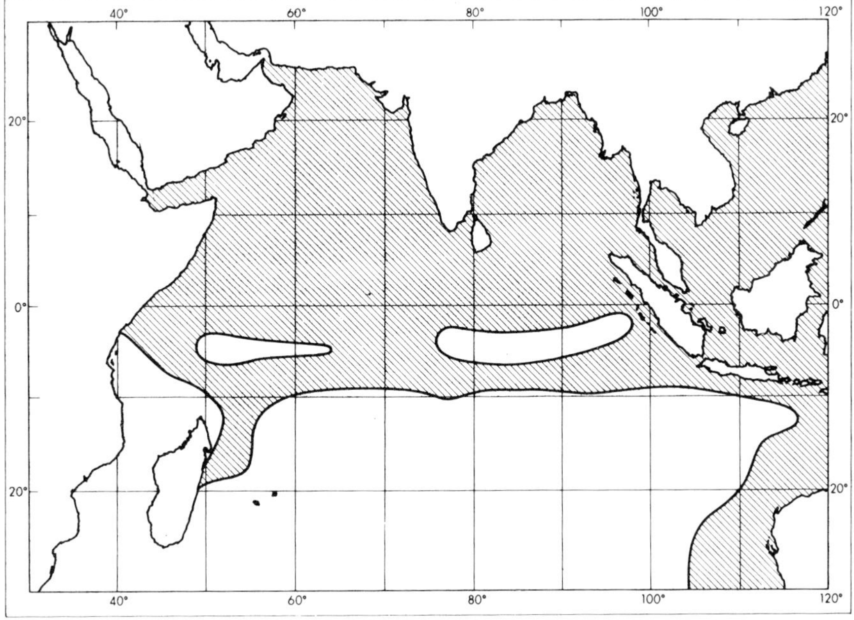

FIGURE 4: Surface current direction as criterion for the monsoonal area. A change in the mean resultant surface current direction of more than 90° (hatched area) indicates that the ocean circulation possesses a monsoonal character. The evaluation is based on ship drift observations.

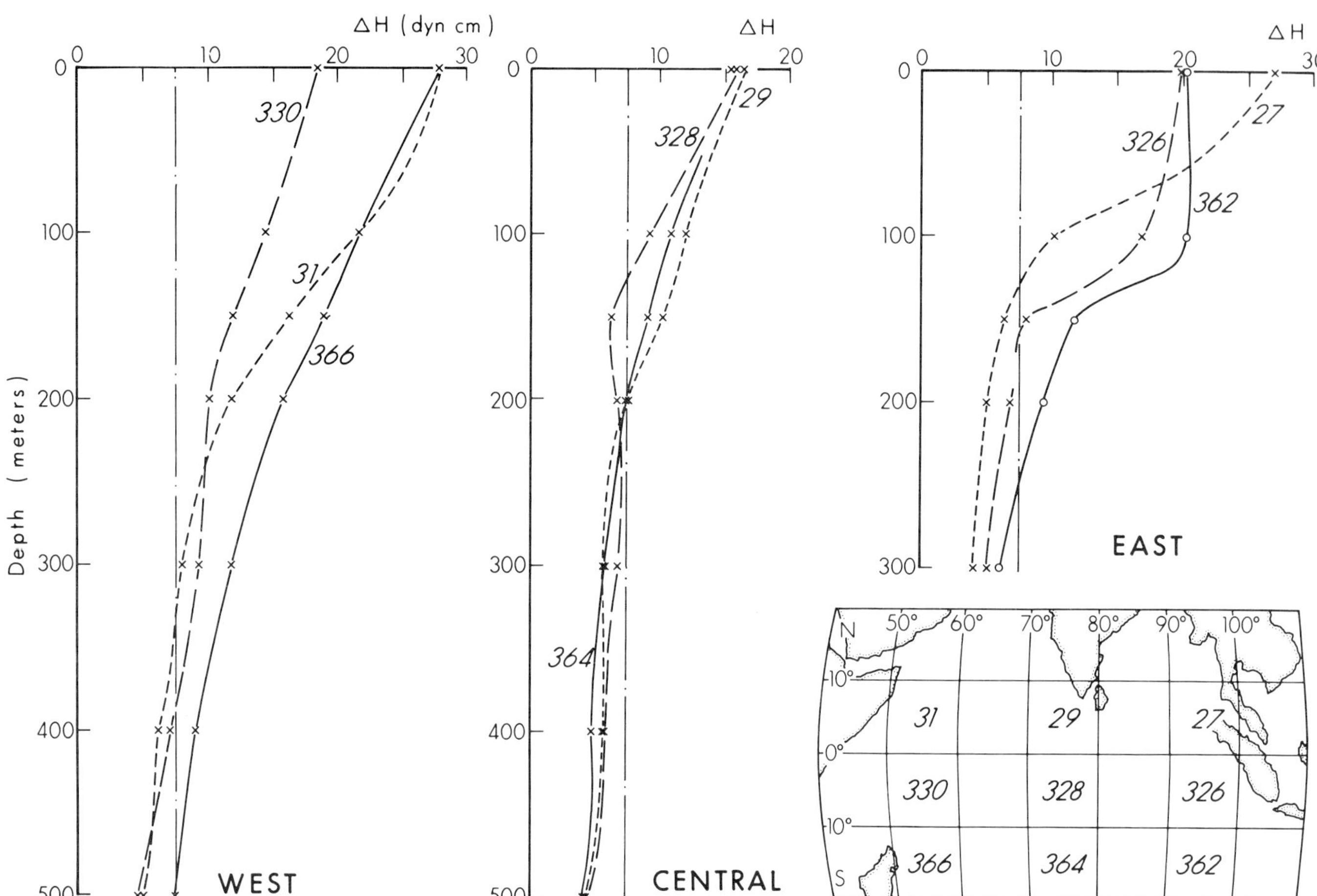

FIGURE 5: The vertical distribution of △ H for the western, central, and eastern regions. Numbers given are Marsden squares.

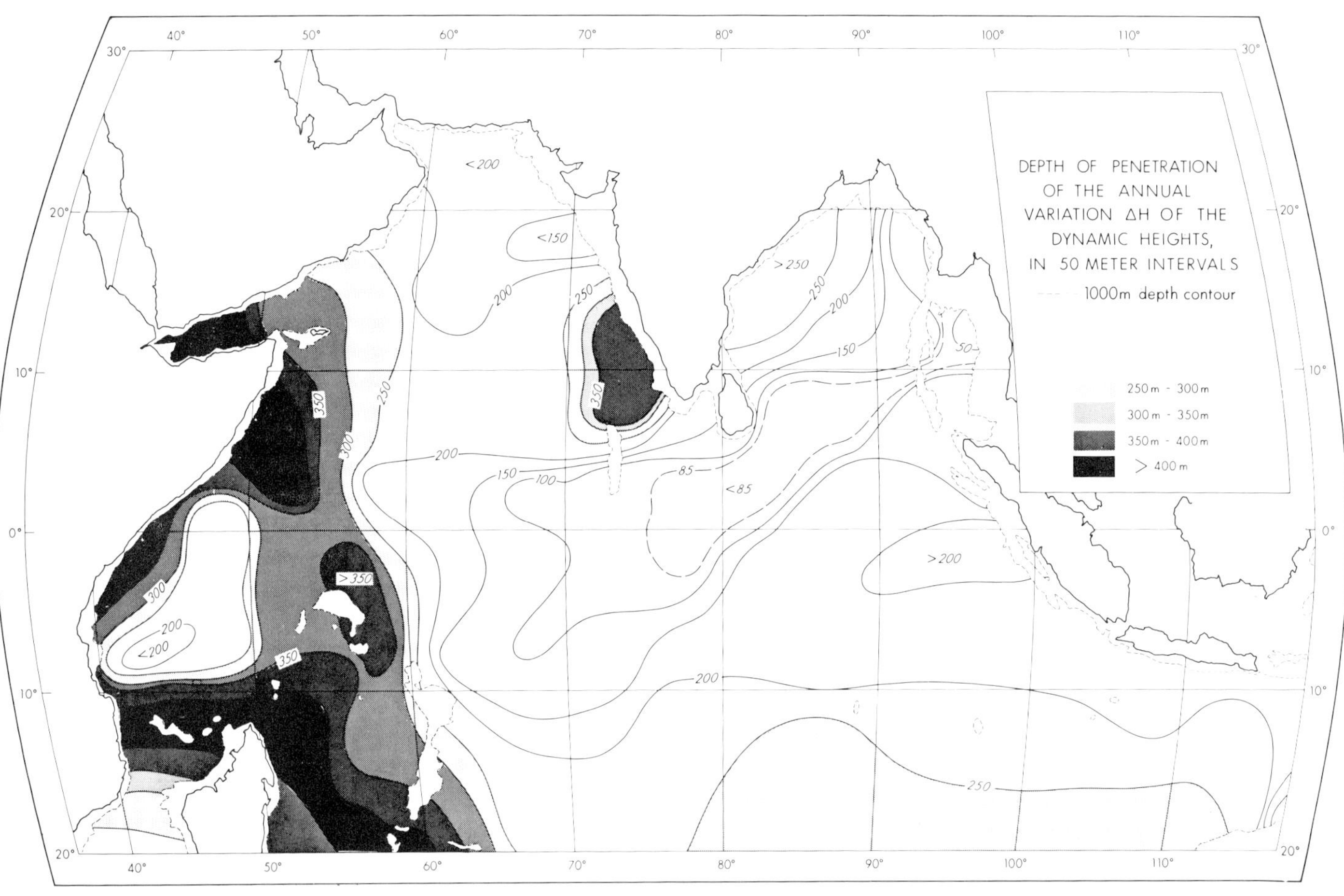

FIGURE 6: A definition of the vertical extension of monsoonal effects.

able annual variation. One recognizes from Figure 5 that this choice of a lower-limit value ΔH — as indicated by the dashed vertical line — implies a certain degree of arbitrariness; this, however, hardly influences the relative distribution of penetration depths.

When studying the chart of penetration (Fig. 6), one must keep in mind that the effects caused by the thermohaline circulation cannot be separated from the effects caused by the wind. This is especially true in the Arabian Sea: Düing and Schwill (1967) have shown that the highly saline water from the Persian Gulf at the level between 200 m and 300 m undergoes considerable seasonal fluctuations. The over-all pattern of the chart of penetration of ΔH reveals much higher values for the western part of the ocean than for other areas. A sharp decrease occurs along the longitude of 60° E. A minimum is found in the central parts of the northern Indian Ocean, a relative increase of the penetration depth, along the eastern boundaries.

The discussion here is not intended to give an explanation for the pattern displayed in Figure 6, for the many different factors and their interactions, such as thermohaline and wind-driven circulation, advective influx of water masses, and effects of westward intensification, make such an attempt extremely difficult and most doubtful. Figure 6, therefore, should be considered only as an attempt to give a coarse definition of the depth penetration of the monsoonal circulation.

C. Characteristic Periods In the Northern Indian Ocean

During the earlier phases of processing the data of the International Indian Ocean Expedition, there was hope that it would be possible to draw monthly charts for the physical parameters of the Indian Ocean. It soon became obvious, however, that because the available data were so heterogeneously distributed in time and space a compromise — combining the data into several periods — had to be found. The decisions made were based on three different atlases: *Monatskarten für den Indischen Ozean* (Deutsches Hydrographisches Institut, 1960); *Atlas of Surface Currents Indian Ocean* (Hydrographic Office, 1958); and *Atlas on General Air Circulation Indian Ocean* (Koninklijk Nederlands Meteorologisch Institut, 1952). The different atlases indicate that the spring transition for the atmospheric circulation occurs on the average during April and the fall transition, during October–November. It is known, however, that the transition in certain years can occur earlier or later. Furthermore, the transitions can occur at different times from region to region in any given year. In order to get a more accurate determination of the transition periods, therefore, the information contained in the atlases was evaluated by having the seasonal variation determined for the average values of current speeds (based on ship-drift observations) and wind speeds for the western and for the eastern region (Fig. 7). The western region covers the areas 8° N to 4° N (Fig. 7a) and 4° N to 2° S (Fig. 7b) in a zone approximately 5 degrees wide, parallel to the African coast; the eastern region covers the areas 7° N to 3° N (Fig. 7c) to 3° N to 9° S (Fig. 7d) in a 5-degree-wide zone centered around 92.5° E.

A comparison of the graphs for the western and eastern regions shows that the amplitudes for both wind and current speed are considerably higher in the west and that the summer period lasts longer

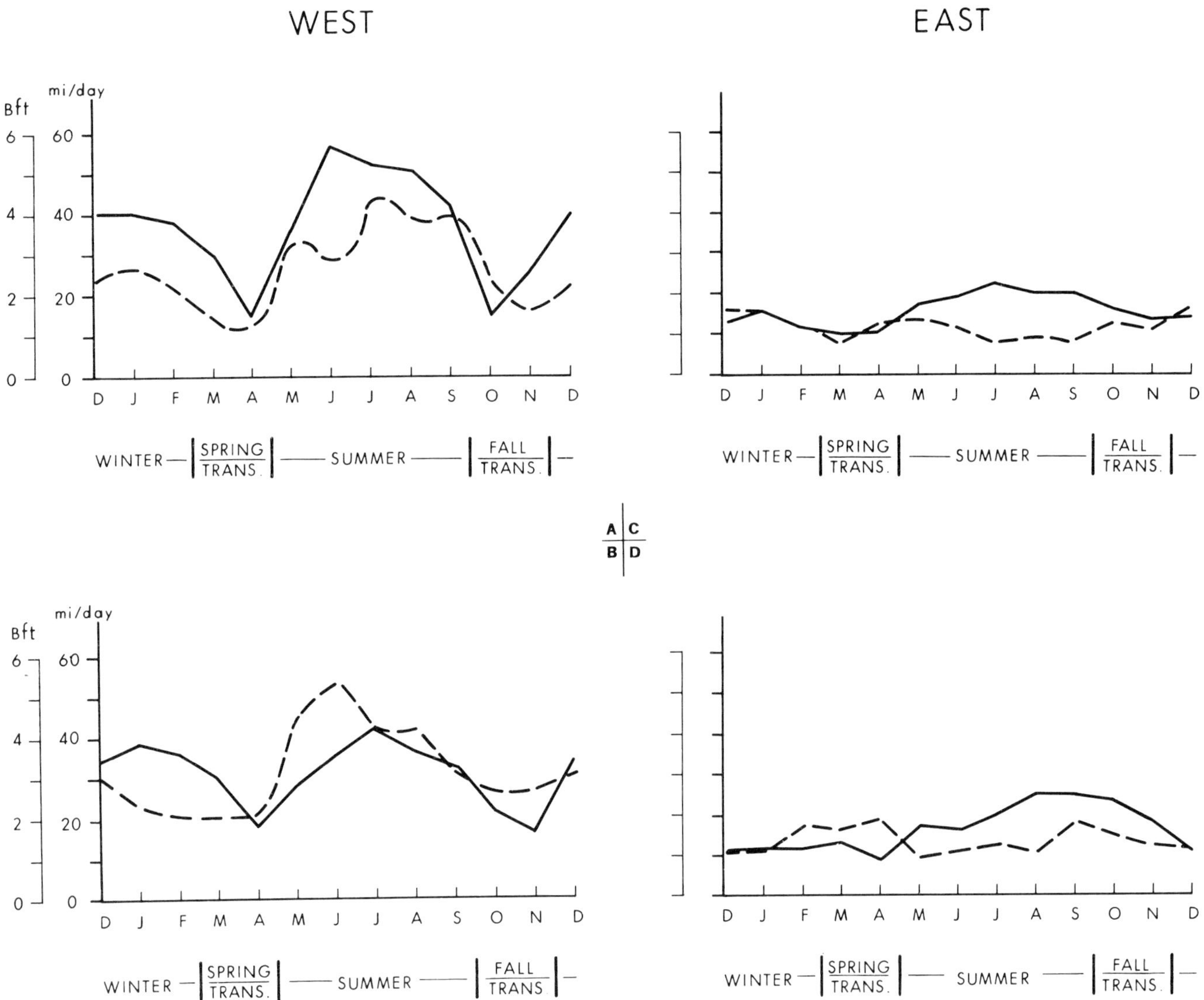

FIGURE 7: Seasonal variation of the average values of current speeds (— — — — —) and wind speeds (—————) in the western and eastern region of the northern Indian Ocean. For exact locations see text.

than the winter period. Thus, March–April and October–November have been chosen to represent the two transition periods. (This combination of two months was caused partly by the lack of data.) Hence, the summer period is considered to extend over five months, from May to September; the winter period, over only three months, from December to February. Moreover, to reveal the rapid changes in the circulation pattern, the summer period is split into two parts: May–June and July–August–September.

Unfortunately, it does not seem to be possible, from the use of the information contained in the atlases, to give any reliable answers to the important question of the phase lag between wind and surface current.

II. PREPARATION OF THE DYNAMIC TOPOGRAPHY MAPS

A. Choice of the Reference Level

The choice of the 1000 dbar level as the reference level was based on the following considerations: the dynamic-height anomalies relative to 3000 dbars were computed for the levels of 2000, 1000, 800, and 600 dbars; these computations were based on 721 stations reaching 3000 m or greater depths; the area covered lies north of 20° S; and all stations that were made during the entire period under observation were used.

The distribution of dynamic-height values for the 2000 and 1000 dbar levels is almost completely random, and no appreciable horizontal gradients occur. Figure 8a represents the frequency distribution of the dynamic-height values for the 1000 dbar level relative to 3000 dbar. The graph shows that 84% of the values have a standard deviation of ± 5 dyn cm or less. For comparison, Figure 8b gives the distribution for the 600 dbar level relative to 3000 dbar. In this case, only 68% of the values have a standard deviation of ± 5 dyn cm or less, which indicates that noticeable horizontal gradients are involved. Again, for comparison, an extreme case is presented in Figure 8c: the frequency distribution for the dynamic-height values from July, August, and September for the sea surface relative to the 1000 dbar level does not show a narrow Gaussian distribution, owing to the large horizontal gradients involved.

It may be noted here that Swallow and Bruce (1966) used the same reference level in an investigation of the currents near the Somali coast during the summer monsoon.

B. Treatment of the Dynamic-Height Values

Although the material brought together at the University of Hawaii is the most comprehensive data on the Indian Ocean, it is still very heterogeneous for the following reasons:

1. Even the stations made during one cruise are not synoptic; they are separated by weeks or months. Thus, in order to cover each period, data from many cruises made in different years have been combined.
2. The existence of internal waves can produce horizontal pressure gradients that easily lead to misinterpretations. A striking example was given by Defant (1950).

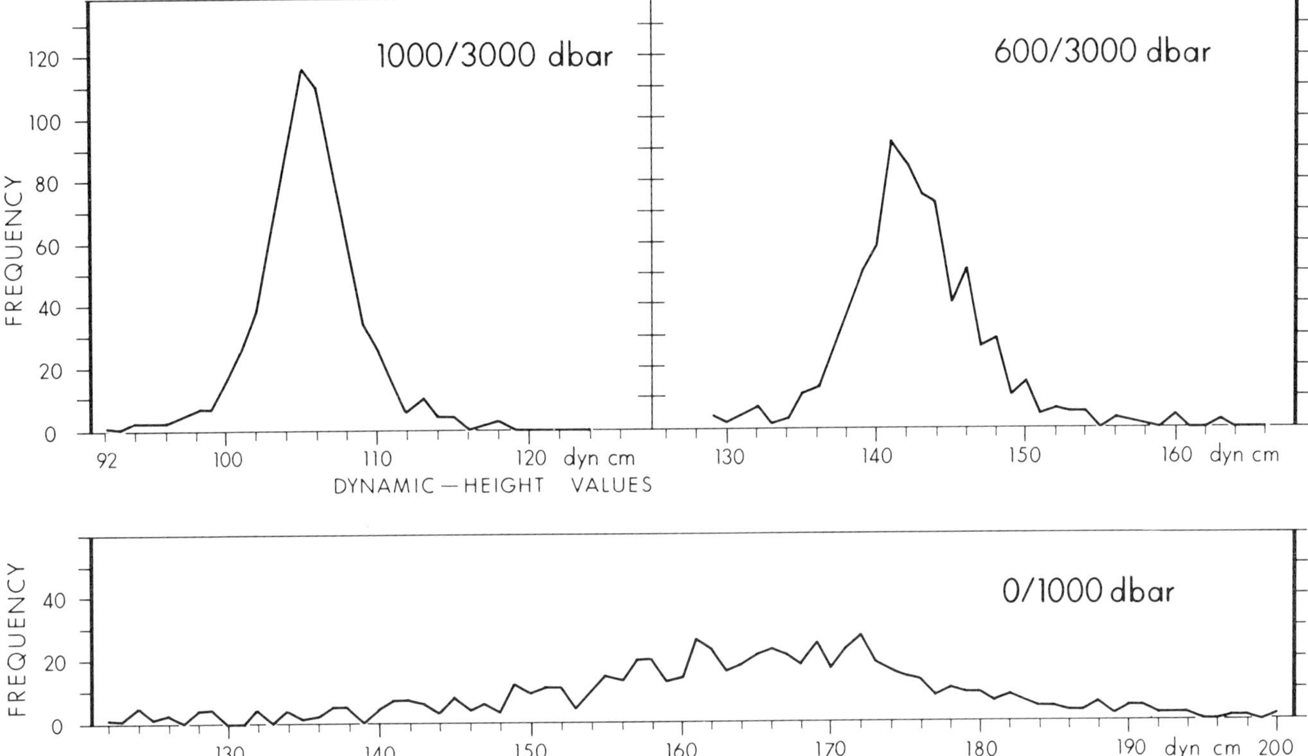

FIGURE 8: Frequency distribution for the dynamic-height values in the Indian Ocean north of 20° S.

(a) 1000 dbar level referred to the 3000 dbar level using data from the whole year.

(b) 600 dbar level referred to the 3000 dbar level using data from the whole year.

(c) 0 dbar level referred to the 1000 dbar level using data from July, August, and September only.

3. Because the observations were carried out by sixty-two research vessels from seventeen nations, it may be expected that the standards of observations and data reduction were not uniform. This nonuniformity, combined with errors in the determination of the temperature, salinity, and depth, affects the computed values of the dynamic heights.
4. As was pointed out before, the northern part of the Indian Ocean is characterized by its nonstationary behavior. Thus variations that in reality take place over a time span of several months obscure the presentation of the mean picture for one period.

By choosing the reference level at 1000 dbars, the number of stations was reduced from a total of 10,037 to 4,823. These stations cover the whole Indian Ocean, with the exception of the Red Sea and the Persian Gulf, about 70% of which are distributed over the area north of 20° S.

In order to remove some of the heterogeneities, the following procedure was applied: For each 5-degree square, the mean and the standard deviation of all dynamic-height values for the sea surface from one period were computed. The difference between the mean value and the value of each individual station was then determined. If this difference exceeded one- and one-half times the standard deviation, the corresponding station was discarded. To gain an insight into the accuracy of the dynamic-height values used, the over-all mean values and standard deviations of the total area north of 20° S have been computed. Their seasonal distribution is given in Table 1.

One should keep in mind, however, that a standard deviation computed over a 5-degree square not only may contain erroneous values but also be influenced by horizontal variations of the dynamic topography that are real. This is especially true for the western part of the ocean, where high gradients occur. In practice, it seems that a contouring interval of 5 dyn cm is well justified.

In applying the described procedure, 433 of the 4,823 stations were rejected.

Table 2 gives the monthly distribution of the remaining 4,390 stations for the entire Indian Ocean.

TABLE 1:   **Over-all Mean Values and Standard Deviations for the Area North of 20° S (in dyn cm)**

Period	Over-all Mean Values	Over-all Standard Deviation
Winter	166.2	±6.0
Spring	167.0	±5.3
Early summer	161.1	±5.8
Late summer	165.0	±6.0
Fall	164.0	±6.6

TABLE 2:   Number of Stations during Different Months and Seasons

Month	Number	Number	Season
March	538		
April	486	1024	Spring
May	446		
June	320	766	Early summer
July	299		
August	394	1029	Late summer
September	336		
October	349		
November	321	670	Fall
December	269		
January	331	901	Winter
February	301		

It is of interest to compare the figures in Table 2 regarding the number of stations used with those of similar investigations for other oceans: Defant (1941) used a total of 629 stations to contour the dynamic topography of the Atlantic Ocean, and Reid (1961) used a total of 1,792 stations to contour the topography of the Pacific Ocean.

III.   THE DYNAMIC TOPOGRAPHIES OF THE SEA SURFACE

The discussion in this section is not intended to give an extensive description of the dynamic topographies and their variation during the year; only the most striking features will be discussed. The reader may examine the details of the charts given in Figures 9 to 13.

The most outstanding feature common to all charts is the occurrence of a complex pattern of lows and highs representing cyclonic and anticyclonic vortices. Even if one takes into account the fact that the data are of a very heterogeneous quality, there can be little doubt that this complex distribution really exists. In order to substantiate this far-reaching statement, two examples will be discussed in detail: First, only those observations made during July through September 1963 have been used; second, only those made from May and June 1964 have been used.

The first example concerns the dynamic topography for the Arabian Sea during summer 1963, based on observations made by the research vessels, *Atlantis II, Discovery,* and *Anton Bruun.* As was to be expected, the dynamic topography based on the data obtained from a single year, 1963 (Figs. 14, 15), shows substantial deviations from the topography that is based on data obtained from several years (Fig. 11). This is caused mainly by the discarding of stations that have extremely high or low values — the discarding procedure has been described in the previous section. The pattern of changing highs and lows of the sea level (Fig. 14) that extends eastward from the Somali coast is based essentially on two *Atlantis II* sections made during August and September 1963.

The topographies for the 100, 200, 300, 400, and 500 dbar levels, based on the same data, were also contoured; but they are not shown in this study. It can be said, however, that the cell pattern is still clearly present at the 400 dbar level. In addition, the geostrophic mass transports from these two *Atlantis II* sections, as computed by Bruce (1968), are given in Figure 16. The vortex pattern (Fig. 14) fits well into the general picture of the dynamic topographies of the northern Indian Ocean. Its regularity, however, raises the question of whether it is caused wholly or partly by internal waves. In an attempt to make at least a rough check on this possibility, the dynamic heights as computed from five different *Atlantis II* sections in the northwestern part of the Indian Ocean were plotted as a function of the observation time (Fig. 17).

In connection with the present problem, periods in the range of half-daily and daily tides, as well as in the inertial range, are of interest. The inertial periods are: for $15°$ N $= 1.9$ days; for $10°$ N $= 2.9$ days; and for $5°$ N $= 5.8$ days. According to Dietrich (1957), half-daily tides are predominant in the Arabian Sea. Not one of the five sections in Figure 17 indicates the existence of either half-daily or daily periods. In four of the sections, the inertial periods of the corresponding latitudes seem not to be involved. One possible exception is the *Atlantis II* section along $10°$ N made during summer 1963, where the inertial period of 2.9 days possibly has a certain influence. A strong argument against this possibility, however, is the fact that the cell structure reaches down to the 400 m level. If this cell structure were simulated by internal waves, it would imply extremely great amplitudes. The existence of internal waves of such great amplitudes seems to be unlikely for the upper layer of a tropical ocean that has a very high density gradient. It seems possible, however, that internal waves of considerable amplitudes are generated in deeper layers, where the vertical density gradients are weaker. Assuming that such deep internal waves do not affect the upper 200 m of the ocean, one should expect that a surface topography based on the 200 dbar reference level is also free of disturbances created by internal waves. This expectation was tested on the basis of the data from summer 1963 (Fig. 15). The surface topography based on the 200 dbar reference level shows almost the same pattern of vortices as the topography based on the 800 dbar level (Fig. 14).

As a result of these estimates, it can be concluded that internal waves most probably are not the cause for the cell-like structure in the dynamic topography in Figure 14, but that this structure is a characteristic feature of the oceanic circulation as driven by the monsoon winds. An attempt to give a theoretical explanation of the changing pattern of highs and lows, as well as an interpretation of the exceedingly high figures of mass transport given by Bruce (Fig. 16), will be made in Part 2, Section IV.

For the second example, we shall examine the topography in the May–June chart for the three 10-degree squares from the equator to 10° N and from 50° E to 80° E. In this case, the suspicion could arise that the contoured distribution of high, low, high, low between the Somali coast and the tip of South India and Ceylon is due to observations having been made during the same months (May, June, Fig. 10) in different years. Fortunately, the greater part of the observations in these three 10-degree squares was carried out during the year 1964. The stations made during May 1964 are marked by a circle; those made during June 1964, by a square; and those not marked (in these three 10-degree squares) were made in different years. It turns out that there are 27 stations made during May 1964 and 14 stations, during June 1964. The high and low between 58° E and 70° E can essentially be contoured by using these 41 stations. The high near the Somali coast is based on only a few stations from other years. We know, however, that high gradients (low inshore, high offshore) are characteristic for the Somali Current during summer in general, so that there can be little doubt about the existence of the contoured high near the Somali coast. From these considerations it can be concluded that the pattern of high, low, high from 50° E to 70° E represents a realistic distribution that is not simulated by heterogeneous data. The data in the 10-degree square from 70° E to 80° E have been collected during 1963 and 1965. No conclusions can be made about the reality of the low in this area during May–June of 1964.

A comparison with the sea-surface topographies of the Atlantic Ocean (Defant, 1941) and the Pacific Ocean (Reid, 1961) reveals that the large oceanwide gyres existing in these two oceans are not found in the northern part of the Indian Ocean; and, vice versa, that the conditions prevailing in the northern Indian Ocean are found nowhere else. It appears that the number of vortices is higher during the transition periods than during the summer and winter periods. A comparision of the charts for spring, early summer, and fall with the charts for late summer and winter reveals that the topographies of the first three periods are in general more complex (more highs and lows) than for the latter two. A comparision of the May–June chart with the July–August–September chart provides an example. The May–June chart shows an alternating pattern of cyclonic and anticyclonic vortices ranging in scale from several hundred kilometers up to one thousand kilometers along the zonal belt from the equator to 10° N. The topography for late summer shows a reduced number of vortices of a more elongated shape in the east-west direction.

A further feature common to all charts is an upward slope of the sea surface, from the western to the eastern part of the ocean, that can be observed over a wide range of latitude from approximately 8° S to 20° N. The difference in sea level is especially distinct between the Arabian Sea and the Bay of Bengal. Typical values for the Bay of Bengal are 20 to 30 dyn cm higher than those for the Arabian Sea. This dissimilarity can qualitatively be explained by the different climatic conditions over the western and eastern parts of the Indian Ocean: The western part has an arid climate with a high rate of evaporation, whereas the eastern part has a humid climate with a high rate of precipitation. In addition, the Bay of Bengal and the adjacent areas are affected by the strong river run off. These climatic conditions cause high salinities in the western part and low salinities in the eastern part. This salinity distribution is, of course, reflected in the values for the dynamic heights. The dynamic topographies for the 100 m level, which will be given in the oceanographic atlas of the International Indian Ocean Expedition, indicate that there are no remarkable east-west gradients. It can be concluded, therefore, that this climatic effect is limited to a shallow surface layer of the ocean.

Furthermore, it is of interest to compare the structures of the topography in the Arabian Sea and the Bay of Bengal: The Arabian Sea shows a complex and seasonally changing pattern and, partly, a complete reversal of the surface circulation; in the Bay of Bengal, we find an anticyclonic vortex that appears to be rather consistent throughout the year, although no data are available in the fall. In order to give a qualitative explanation for this striking difference, one can say that the topography of the Arabian Sea is influenced mainly by the monsoons, which in general are stronger in the western part than in the eastern part of the Indian Ocean. In the Bay of Bengal, the dominant thermohaline effects caused by the high precipitation and the high river run off on the eastern side of the Bay bring about the anticyclonic vortex. For further details about the area, see La Violette (1967).

In the remainder of this section, an attempt will be made to give a brief discussion of the dynamic topographies in terms of the conventionally known surface current systems. This description will be limited to the most prominent current systems in the tropical latitudes: the North and South Equatorial Currents, the Countercurrent, the Southwest Monsoon Current, and the western boundary currents that develop near the Somali coast during summer and winter. In view of the complex topographies, it seems appropriate to renew a statement made by G. Schott, as early as 1935, in his excellent book on the geography of the Indian and Pacific Oceans:

Im Anschluss an neuere Auffassungen tut man gut, sich nicht mehr auf geschlossene ozeanische Kreisläufe mit oder gegen den Uhrzeiger festzulegen, obwohl bei den Oberflächenströmen die Neigung dazu offenbar vorhanden ist. Auch sollte man von der Vorstellung abgehen, dass es dauernd unbewegte Teile der Meeresoberfläche gibt, etwa in der Mitte der "Kreisläufe." Doch pflegt in beträchtlichen Flächen eine ausgesprochene Strömung zu fehlen. Es sind sogenannte neutrale Gebiete, wo der Strom auf kurze Entfernungen die Richtung wechselt und seine Geschwindigkeit auf ein Minimum herabgehen kann.

Following modern concepts, it is advisable not to insist on closed oceanic cyclonic or anticyclonic circulation currents, although there is such a tendency. One should also reject the view that the center of gyres are characterized by the permanent absence of currents, although over wide areas of the ocean one does not observe a distinct current pattern. These are "neutral" areas where the current direction changes over short distances and where current velocities reach minimum values. (Author's translation.)

March–April (Fig. 9).   The North Equatorial Current is indicated as a rather narrow belt extending from the equator to 5° N in the western part and to 8° N in the eastern part. The Countercurrent is fairly well developed and extends from equatorial latitudes to approximately 5° S. It appears as a broad band several hundred miles wide in the west, where it seems to be fed by the backflow of the South Equatorial Current. In the eastern part, around 85° E, it narrows to approximately 150 miles in width. The South Equatorial Current is well developed from 100° E to 45° E, where it splits into a northward and southward branch. It extends over a latitudinal range from 8° S to 15° S. With the exception of the months of October and November, the South Equatorial Current seems to be the most consistent current in the northern part of the Indian Ocean apparently because it is located south of the region of monsoonal influence as defined in Figures 3 and 4. The low found to the south of the island of Socotra indicates that there still exists a strong remnant of the winterly Northeast Monsoon Current flowing southwestward along the Somali coast.

May–June (Fig. 10).   The circulation in the area from the equator to 10° N and from 40° E to 80° E has been discussed before. It is virtually impossible to say whether the North Equatorial Current or the Southwest Monsoon Current is present. The Countercurrent seems, however, to broaden and to be absorbed in the late summer into the general eastward drift. The South Equatorial Current is well developed in the western part. The lack of data in the eastern region does not allow any conclusions to be drawn between 75° E and 100° E.

July–August–September (Fig. 11).   With the exception of the westward flow in the two highs centered around the equator between 60° E and 80° E, the Southwest Monsoon drift extends across the whole ocean, from Africa to Sumatra, from 15° N to 7° S. It also includes the powerful Somali Current in the western boundary region. The eastward drift includes the Countercurrent, which cannot be isolated as an individual current. The South Equatorial Current is well developed from 8° S to approximately 18° S in the western part of the ocean.

October–November (Fig. 12).   During fall, the large Southwest Monsoon drift, between the equator and 10° N, is broken down into a low in the western part and a high in the eastern part. The branches that flow eastward are only 100 to 200 miles wide. South of the equator, from 40° E to 70° E, the Countercurrent begins to be separated. The South Equatorial Current does not seem to reach across the ocean from Sumatra to Madagascar. Owing to the lack of data around 60° E to 80° E, contours cannot be considered reliable. The high centered around 9° N and 54° E indicates a strong remnant of the Somali Current.

December–January–February (Fig. 13). The North Equatorial Current is fairly well developed between 8° N and the equatorial latitudes. The Countercurrent has built up further, and there is an indication that it extends from the African coast to 110° E in a band between 4° S and 8° S. Near the African coast, the Countercurrent apparently receives contributions from the South Equatorial Current as well as from the Northeast Monsoon Current, which flows southwestward along the Somali coast. At this time of the year, the South Equatorial Current is again clearly present all across the ocean from approximately 8° S to 15° S.

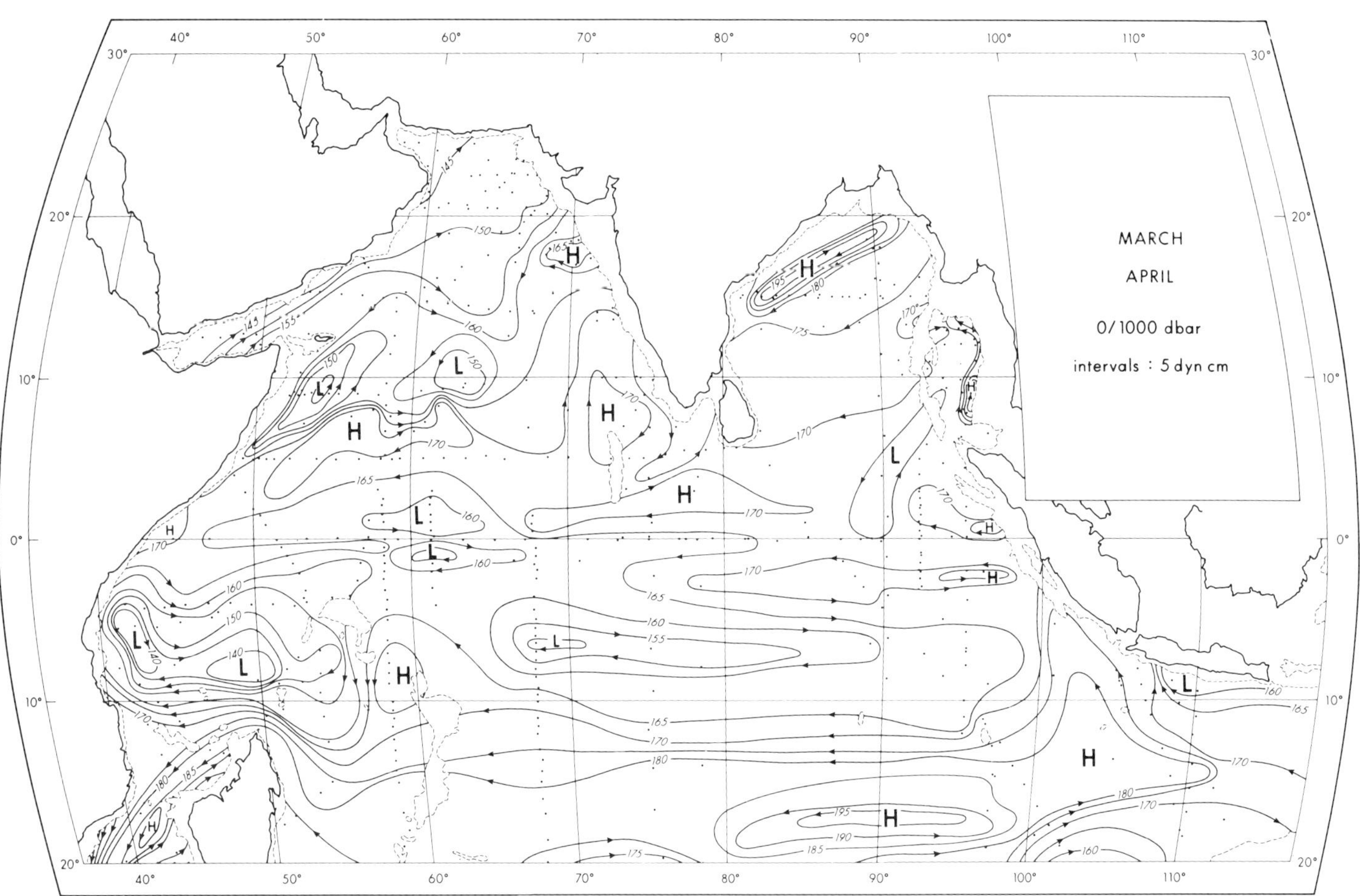

FIGURE 9:   The dynamic topography of the sea surface for spring.

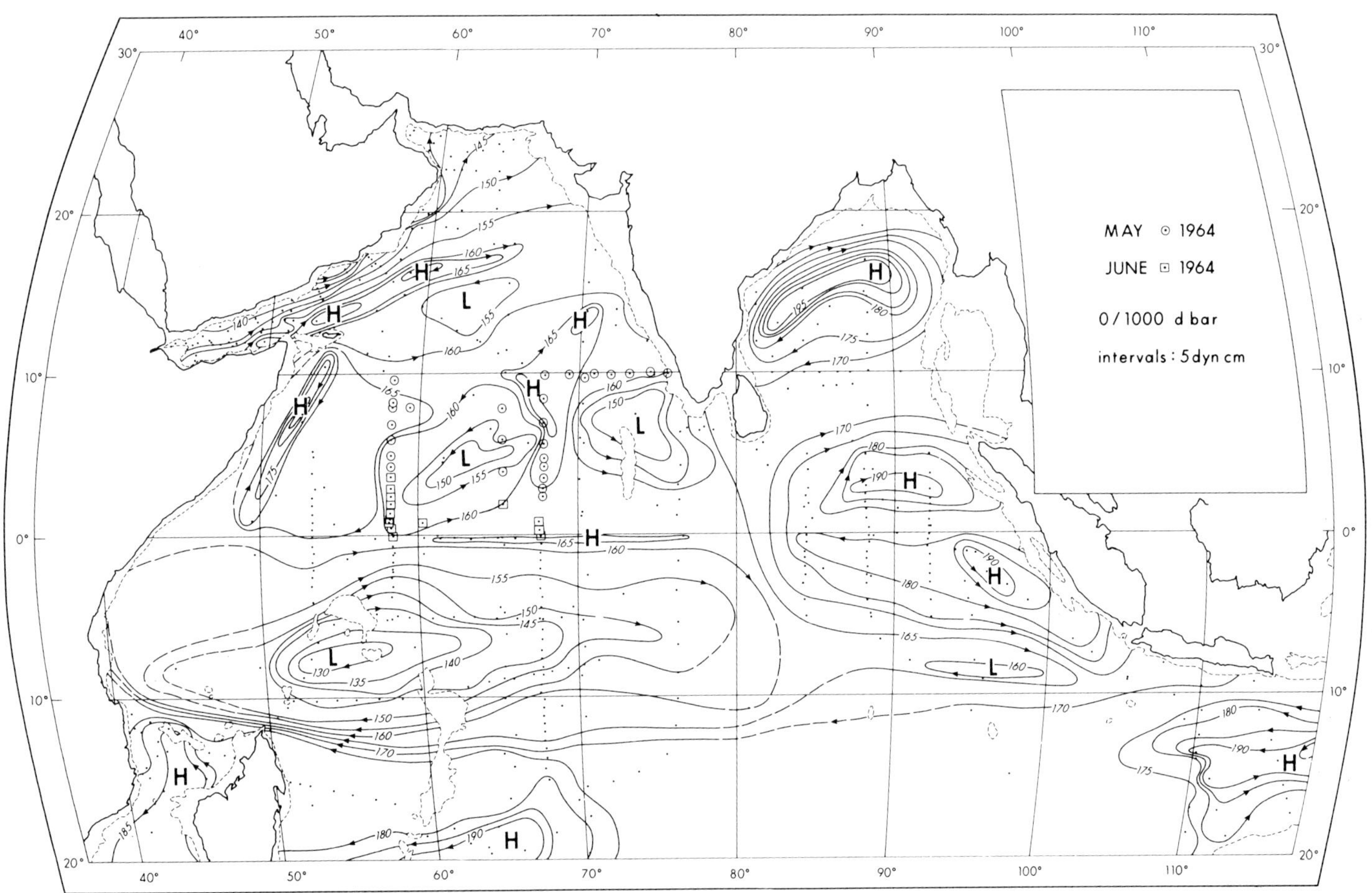

The dynamic topography of the sea surface for early summer.

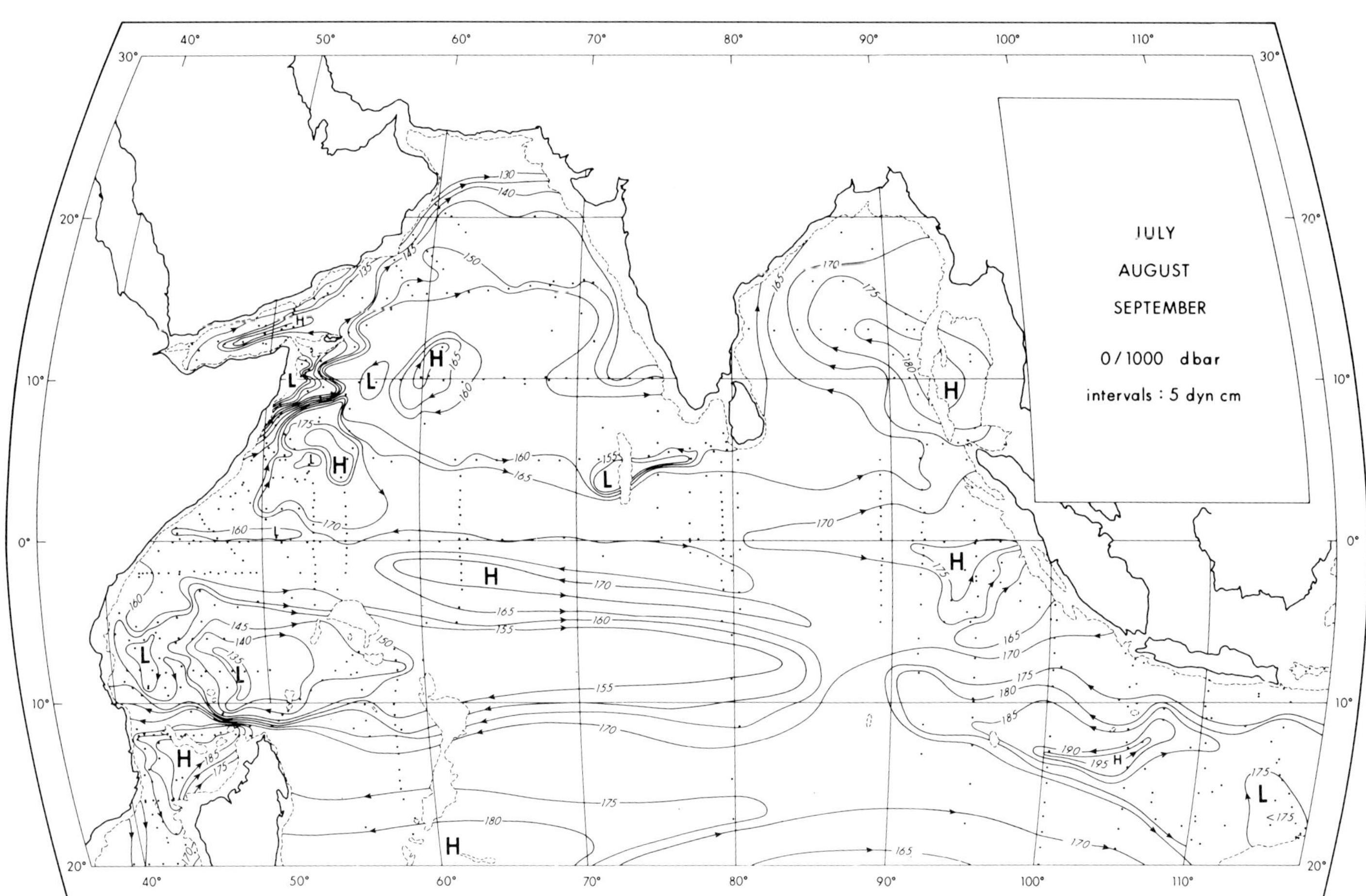

FIGURE 11: The dynamic topography of the sea surface for late summer.

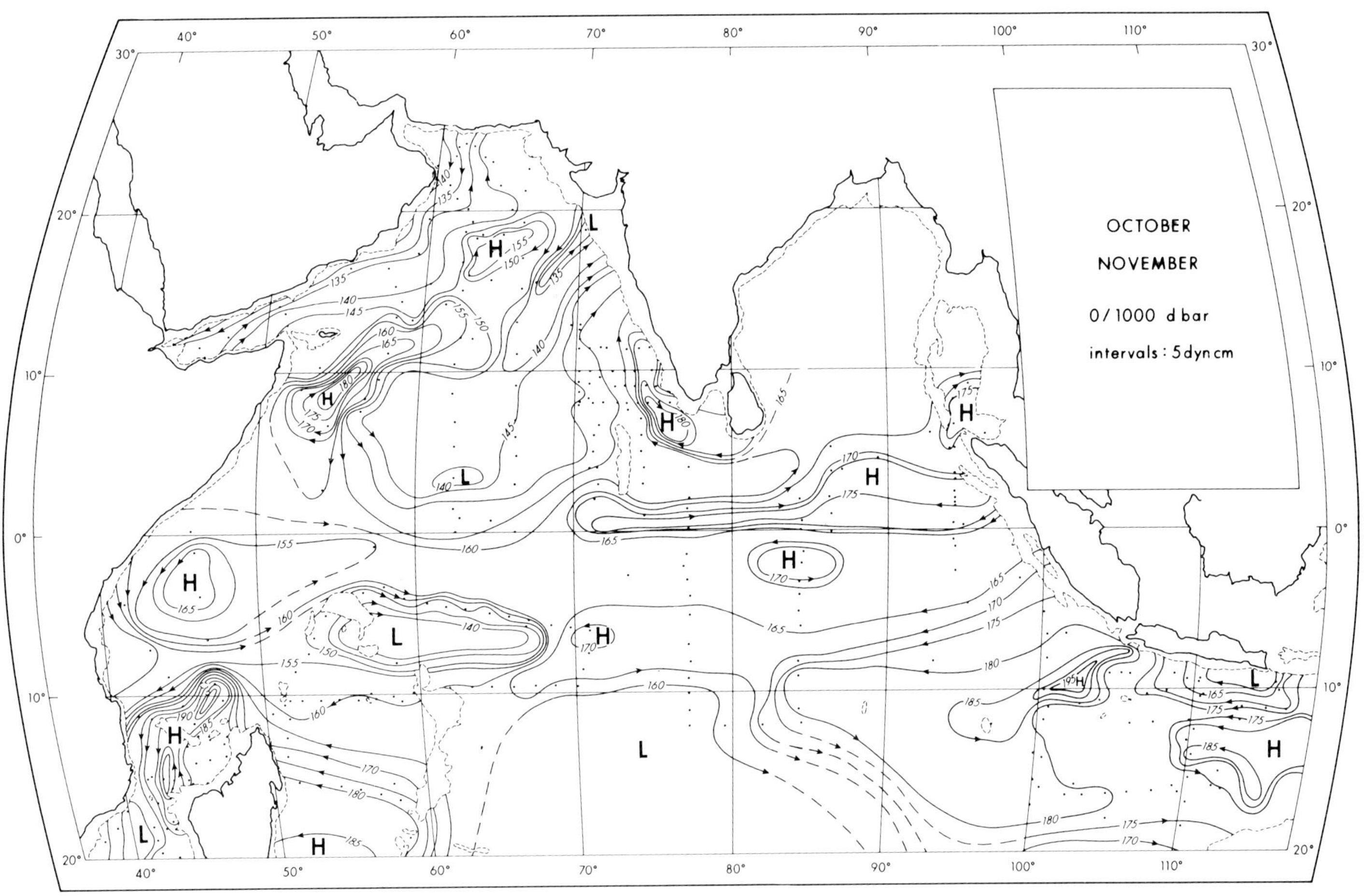

FIGURE 12:   The dynamic topography of the sea surface for autumn.

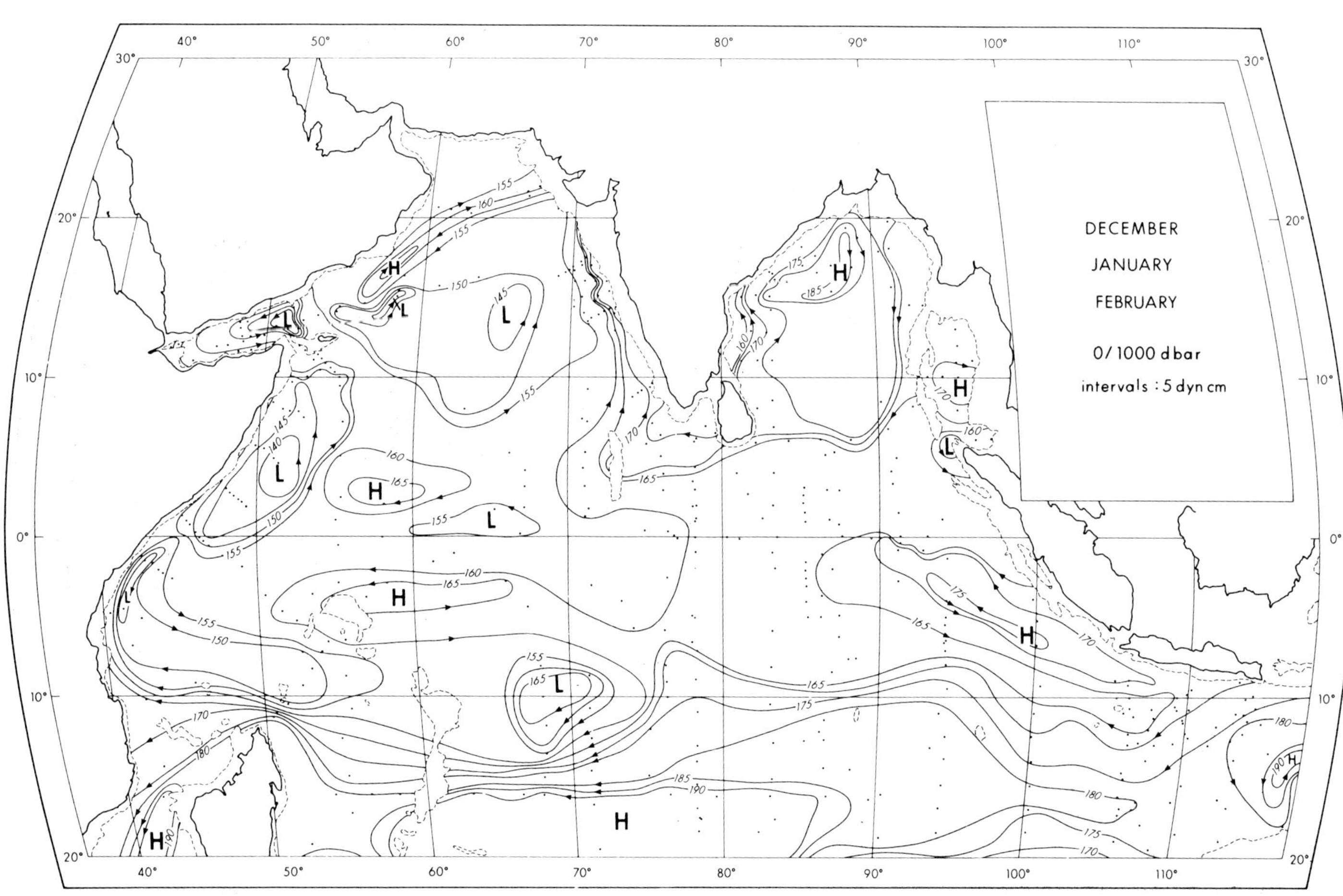

FIGURE 13: The dynamic topography of the sea surface for winter.

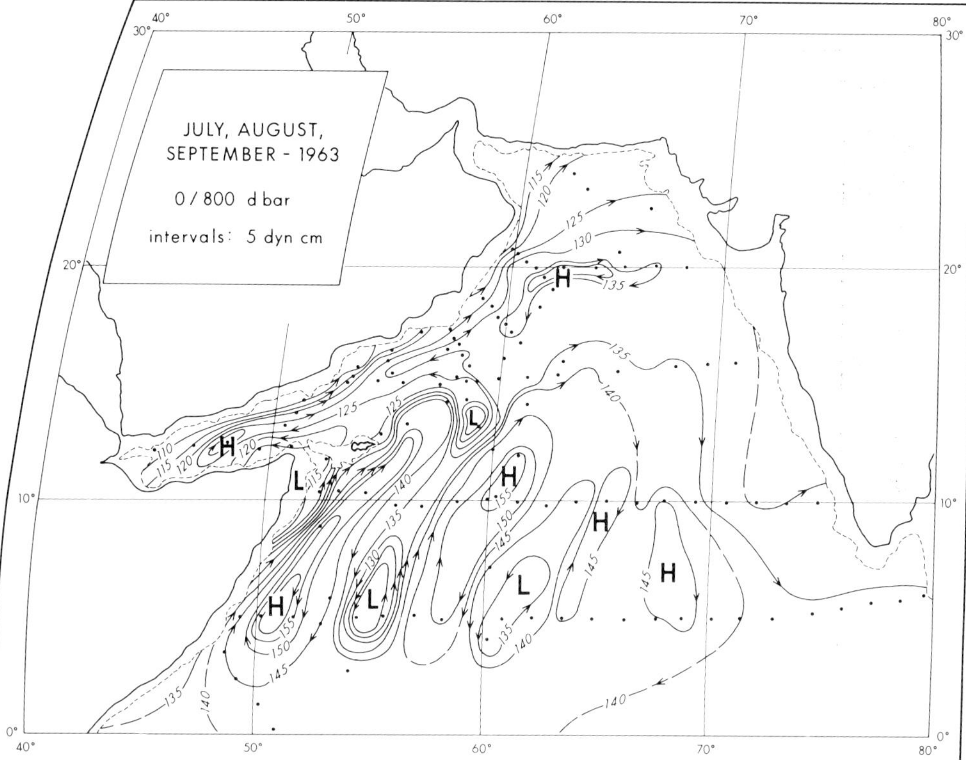

FIGURE 14: The dynamic topography of the sea surface in the Arabian Sea for summer 1963; reference level at 800 dbar.

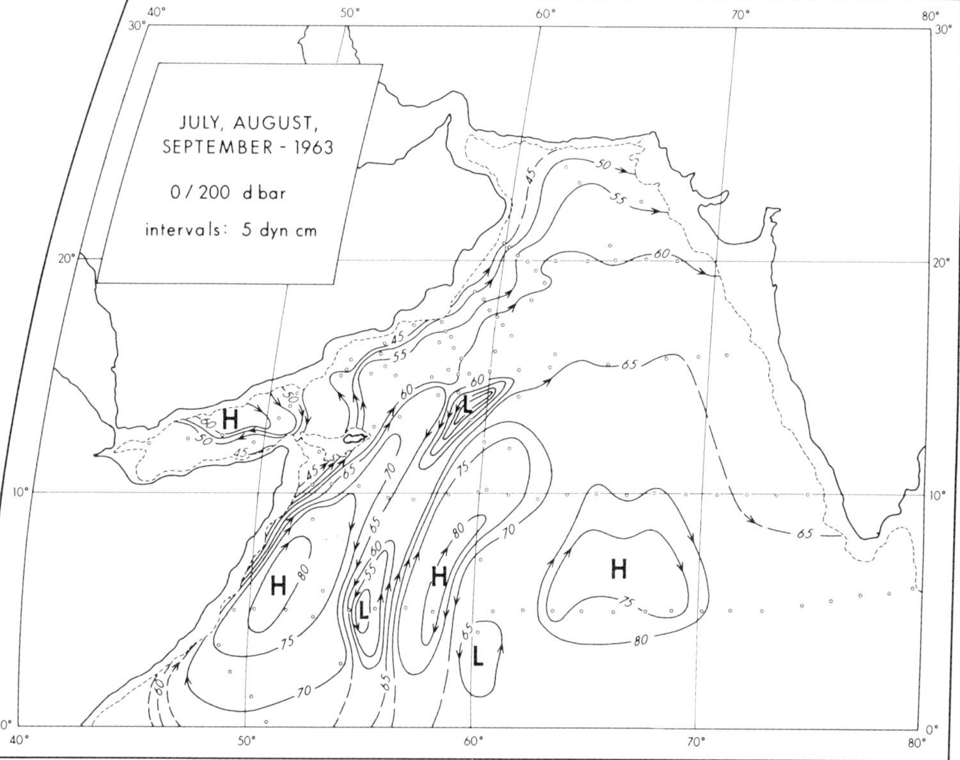

FIGURE 15: The dynamic topography of the sea surface in the Arabian Sea for summer 1963; reference level at 200 dbar.

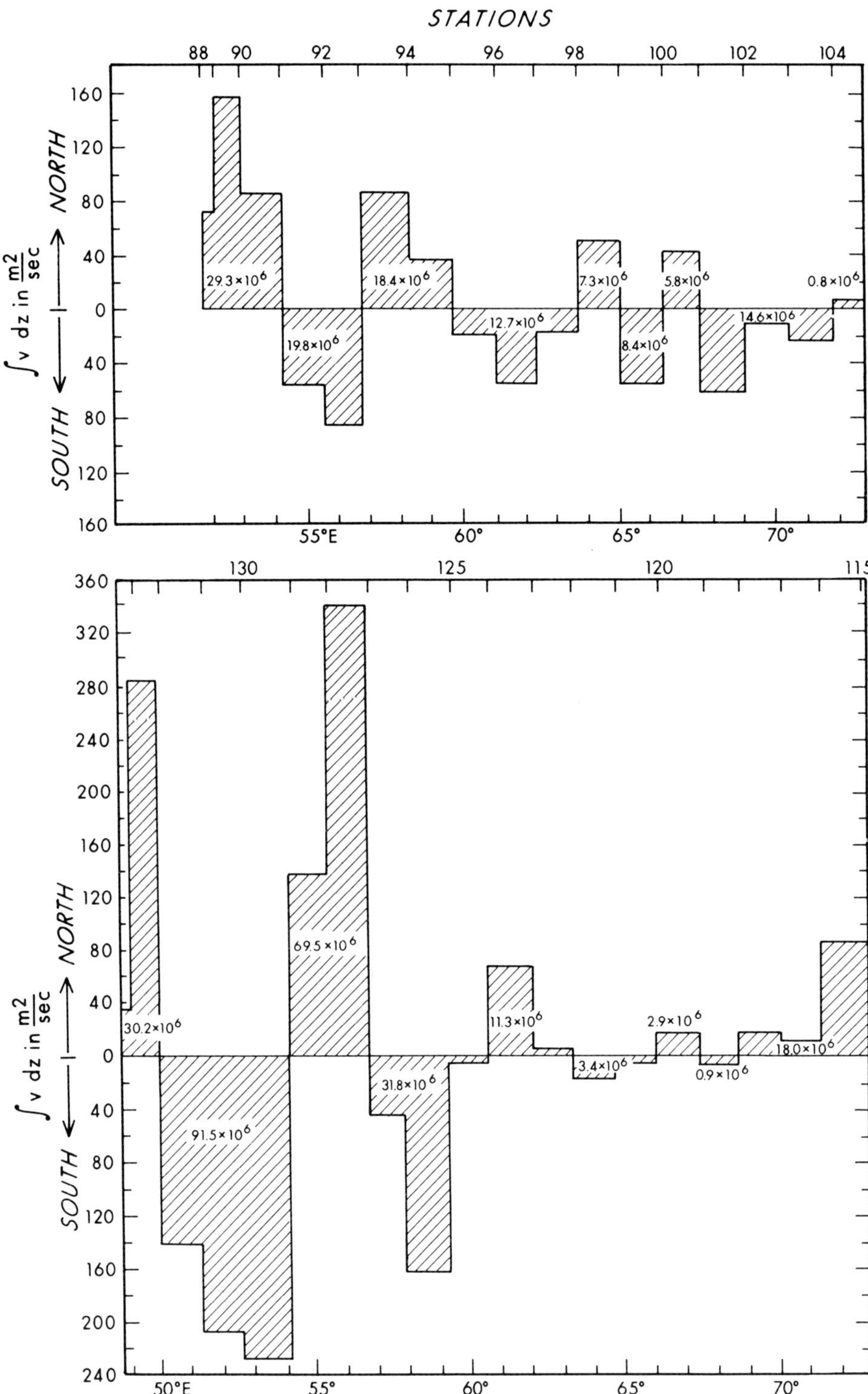

FIGURE 16:   **Mass transport in the meridional direction according to Bruce (1968).**

(a) Section along 10° N using *Atlantis II* stations 88 to 105 as observed during 29 August to 5 September 1963.

(b) Section along 5° N using *Atlantis II* stations 115 to 134 as observed during 13 to 22 September, 1963.

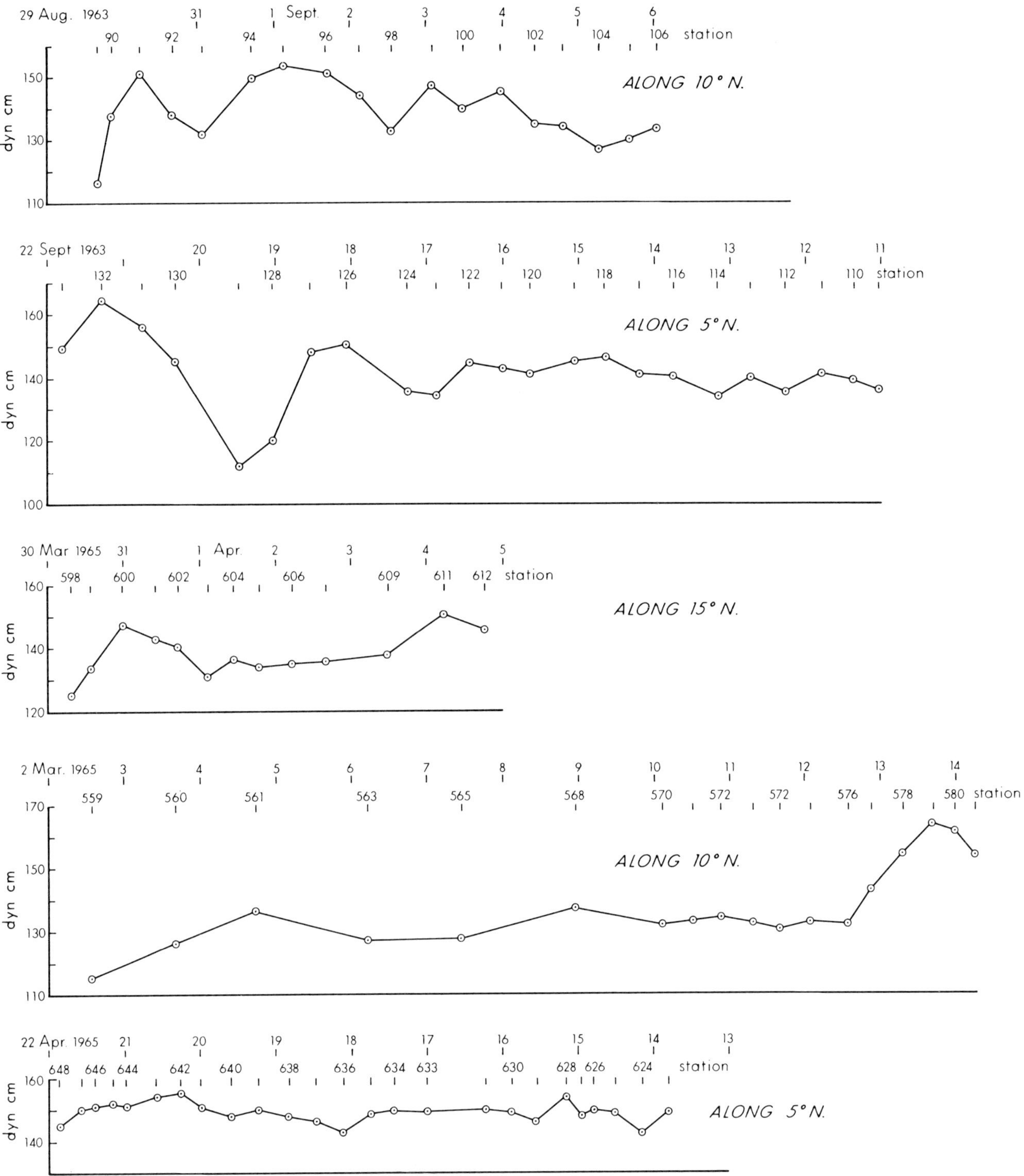

FIGURE 17: The dynamic-height values for the sea surface (reference level 800 dbar) along five zonal sections in the northwestern part of the Indian Ocean. The abscissa gives the times of observations. All observations were made by the research vessel, *Atlantis II*, during 1963 and 1965.

Having discussed the observed circulation, I will now develop a model
of the nonstationary response of the Indian Ocean to the monsoon
winds. The model should give a possible explanation of the highly
complex pattern of circulation represented in the dynamic topographies, especially with regard to the characteristic appearance of
cyclonic and anticyclonic cells. In addition to this more general aim,
the model should give theoretical descriptions of such details as the
phase lag between wind and current and of the conditions during
the transition seasons. One can hope that a nonstationary model de-
scribing the pronounced circulation types of summer and winter in a
satisfactory way will also give a reliable representation of the conditions during the transition periods where observations are lacking
or where nonsynoptic data mask the real pattern.

I. REDUCTION OF WIND OBSERVATIONS INTO ANALYTICAL FORM

The wind distribution over the northern part of the Indian
Ocean is distinguished by two characteristic features that are in
contrast to those found in the other oceans. There is the seasonal
reversal of the wind direction, as well as the strong meridional components associated with the monsoon winds. For the Atlantic and
Pacific Oceans, these meridional components are essentially non-
existent; consequently, they were not taken into account in circulation models. For the Indian Ocean, such an a priori neglect cannot
be allowed.

The vector mean winds for January and July for 5-degree
squares between 10° N and 10° S are given in Figure 18. In July the
winds south of the equator are southeasterly, similar to the southeast
trades of the Atlantic and Pacific Oceans. Near the equator, the wind
becomes progressively less easterly; on the eastern side of the ocean,
it has a westerly component. On the western side of the ocean, the
mean wind does not develop a westerly component until it crosses
the equator. During January the mean wind near the African coast
has an easterly component, whereas, south of the equator, east of
60° E, westerly components prevail.

Two important features can be observed: the wind speeds during
the Southwest Monsoon are generally higher than during the North-

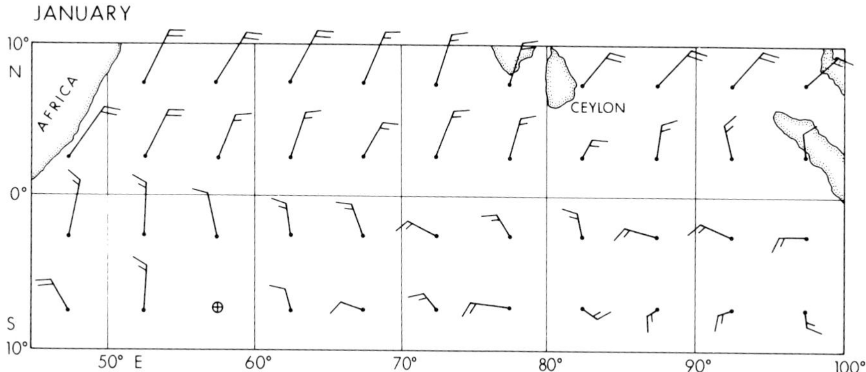

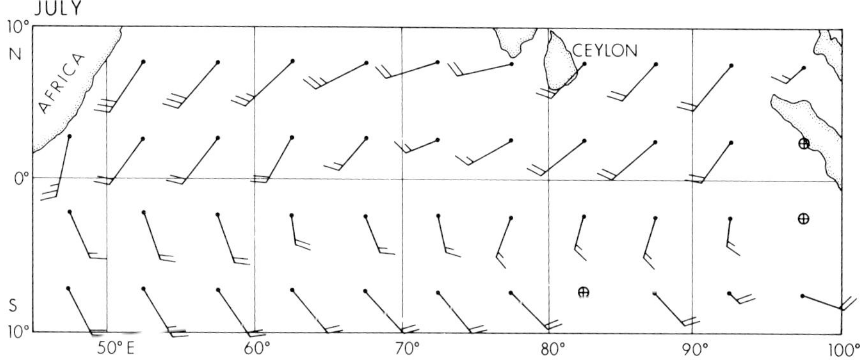

FIGURE 18: Idealized wind observations in the northern Indian Ocean for January and July. After *Monatskarten für den Indischen Ozean*, Deutsches Hydrographisches Institut (1960).

east Monsoon, and the wind speeds during both seasons decrease from west to east. In order to find analytical expressions for the wind observations, the zonal and meridional components of the wind stress have been computed; they are represented in Figure 19.

For the computation of the wind stress, τ, from the observed wind speeds, u, a quadratic relationship of the form

$$\tau = C_D \rho_{\mathrm{air}} u^2$$

has been used, and the drag coefficient has been selected as $C_D = 0.0026$. The validity of this assumption has been discussed by Munk (1950) and Schmitz (1964). The uncertainty in the choice of the drag coefficient C_D leads to uncertainties for the absolute values of the mass transport. For a first-order approach to the problem, the use of such a simple relationship is considered to be sufficient.

The variation of the wind stress with time is considered to be strictly periodical; thus the expressions for both components in a model ocean of length B and width L become

(a) *zonal component* τ_x:

$$\tau_x = -\tau_{x_0} \cos \frac{\pi y}{L} \, e^{i\omega t} \tag{1}$$

$$\tau_{x_0} = 1 \text{ dyne cm}^{-2}$$

$$L = 2000 \text{ km } (y \text{ points northward})$$

$$\omega = \frac{2\pi}{T} \, ; \quad T = 1 \text{ year },$$

(b) *meridional component* τ_y:

$$\tau_y = \left(-\frac{\Delta\tau_{y_0}}{B} x + 0.8 \right) e^{i\omega t} \tag{2}$$

$$\Delta\tau_{y_0} = 0.7 \text{ dyne cm}^{-2}$$

$$B = 6000 \text{ km } (x \text{ points eastward})$$

$$\omega = \frac{2\pi}{T} \, ; \quad T = 1 \text{ year }.$$

Taking the imaginary part into account:

$$t = T \equiv \text{April} , \quad t = \frac{T}{4} \equiv \text{July, etc.} \tag{2a}$$

taking the real part into account:

$$t = T \equiv \text{July} , \quad t = \frac{T}{4} \equiv \text{October, etc.} \tag{2b}$$

According to a personal communication from Professor Ramage (*see also* Ramage, 1968), the assumption of a linear decrease of the wind stress as given in Figure 19b represents a conservative estimate. The wind speeds during the Southwest Monsoon close to the Somali coast near Cape Guardafui and south of it are extremely high. An exponential decrease towards the east would perhaps represent the conditions in a better way. Owing to the lack of data from this high wind-speed area, such an exponential relationship was not used. For possible effects on the boundary current, see Part 2, Section IV. Figure 20 represents the composed analytical wind field over the model ocean for the month of July. Inverting the arrows would give the distribution for January.

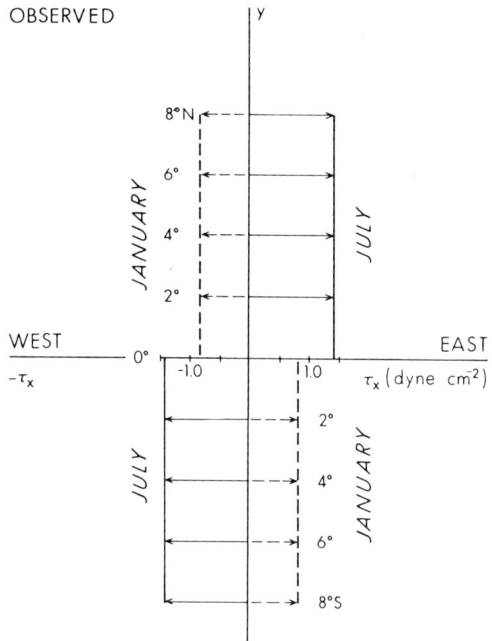

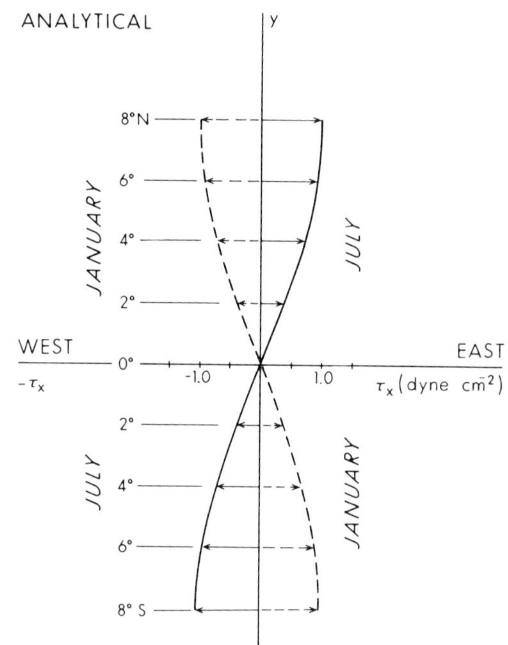

FIGURE 19 *(a):* Zonal components of the wind stress for January and July, according to the idealized observations and the analytical distribution as used in the theoretical model.

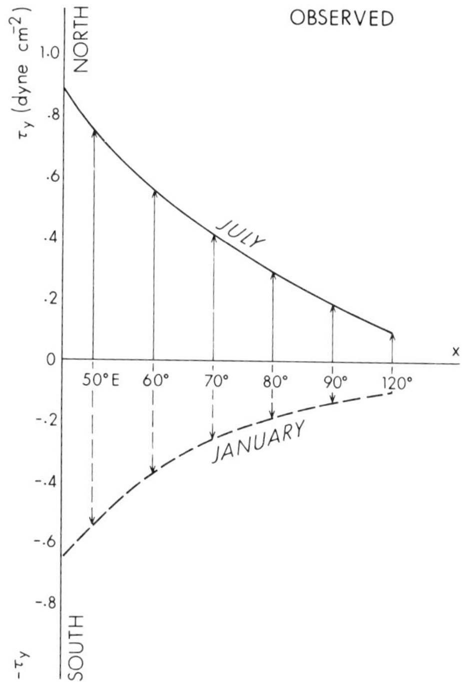

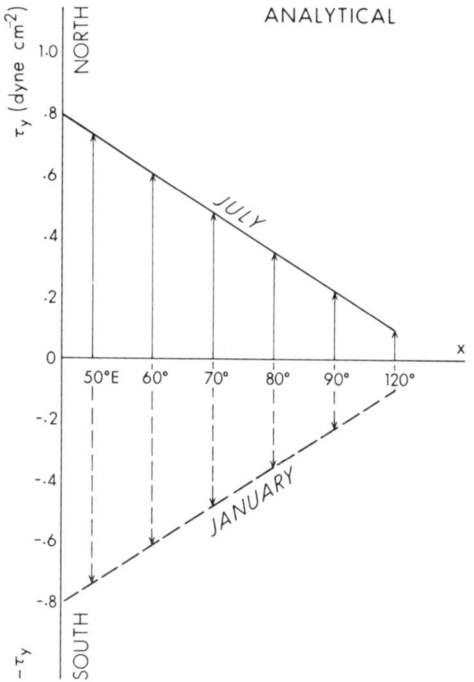

FIGURE 19 *(b):* Meridional components of the wind stress for January and July, according to the idealized observations and the analytical distribution as used in the theoretical model.

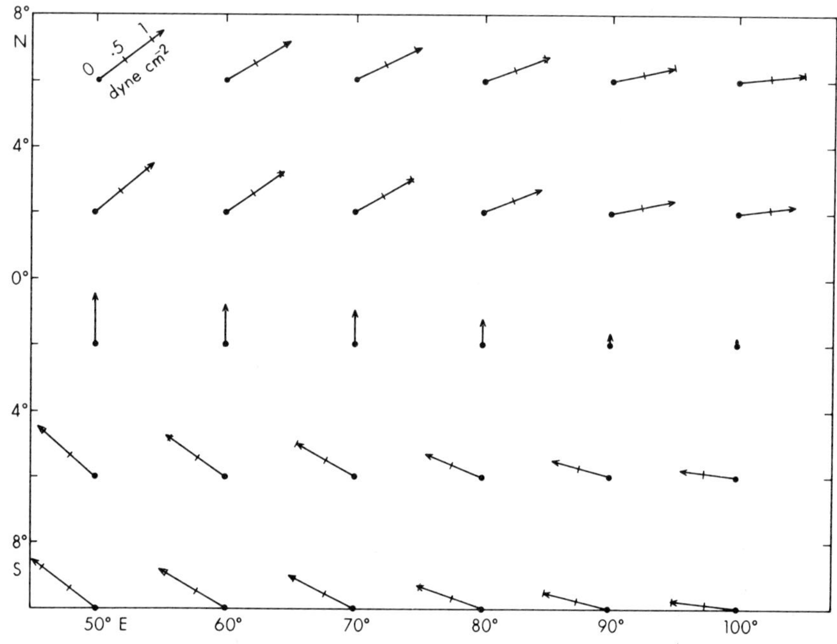

FIGURE 20: Analytical field of wind stress over the model ocean for July. Inverting the arrows would give the distribution for January.

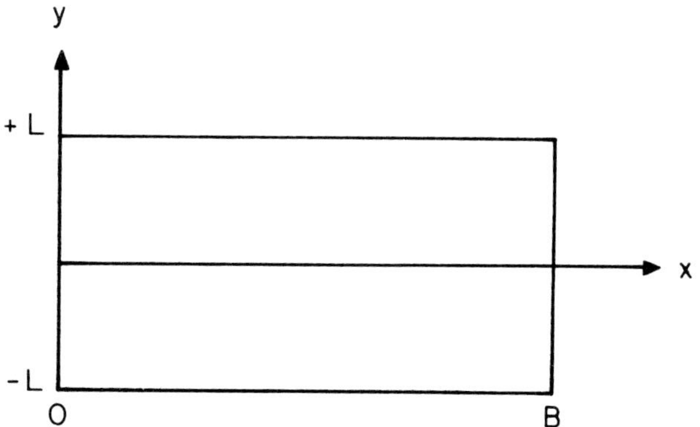

FIGURE 21: The coordinate system for the meridional problem. Note that $L = 1000$ km.

II. FORMULATION OF THE MODEL EQUATIONS

The model ocean might consist of a rectangular basin having a north-south extension of $L = 2000$ km and an east-west extension of $B = 6000$ km. This basin is thought to represent the approximate area from 9° N to 9° S and from 45° E to 100° E.

Let us suppose the ocean surface to be a plane surface at rest: x points eastward, y northward, and z upward. The plane $z = 0$ lies at the mean surface; the actual surface is $z = z_0$.

$$\rho \frac{\partial u}{\partial t} - \rho f v = - \frac{\partial p}{\partial x} + \frac{\partial}{\partial z}\left(k_v \frac{\partial u}{\partial z} \right) - k\rho u \tag{3}$$

$$\rho \frac{\partial v}{\partial t} + \rho f u = - \frac{\partial p}{\partial y} + \frac{\partial}{\partial z}\left(k_v \frac{\partial v}{\partial z} \right) - k\rho v \tag{4}$$

where u and v are the x and y velocity components, p is pressure, f is the Coriolis parameter, ρ is density, and k_v is the vertical coefficient of eddy viscosity. Frictional dissipation is confined to simple bottom friction (bottom of Ekman layer) with a coefficient, k. These equations may be vertically integrated from the surface $z = z_0$ to a depth $z = -h$ beneath which both the currents and the horizontal pressure gradients vanish:

$$\left. \begin{aligned} P &= \int_{-h}^{z_0} p\,dz \\[6pt] M_x &= \int_{-h}^{z_0} \rho u\,dz \\[6pt] M_y &= \int_{-h}^{z_0} \rho v\,dz \\[6pt] \frac{\partial M_x}{\partial t} &= \frac{\partial}{\partial t} \int_{-h}^{z_0} \rho u\,dz \\[6pt] \frac{\partial M_y}{\partial t} &= \frac{\partial}{\partial t} \int_{-h}^{z_0} \rho v\,dz \end{aligned} \right\} \tag{5}$$

The integrals of the pressure gradients are given by:

$$\int_{-h}^{z_0} \frac{\partial p}{\partial x}\,dz = \frac{\partial P}{\partial x} - p(z_0) \frac{\partial z_0}{\partial x}$$

$$\int_{-h}^{z_0} \frac{\partial p}{\partial y}\,dz = \frac{\partial P}{\partial y} - p(z_0) \frac{\partial z_0}{\partial y} \tag{5a}$$

The vertical shearing stress integrates

$$\int_{-h}^{z_0} \frac{\partial}{\partial z}\left(k_v \frac{\partial u}{\partial z} \right) dz = \tau_x(t); \quad \int_{-h}^{z_0} \frac{\partial}{\partial z}\left(k_v \frac{\partial v}{\partial z} \right) dz = \tau_y(t) \tag{5b}$$

where τ_x and τ_y are the components of the wind stress that affect the sea surface. These are time dependent and explicitly given by the expressions (1) and (2). The integrated equations are:

$$\frac{\partial M_x}{\partial t} - f M_y = - \frac{\partial P}{\partial x} - k M_x + \tau_x \tag{6}$$

$$\frac{\partial M_y}{\partial t} + f M_x = - \frac{\partial P}{\partial y} - k M_y + \tau_y . \tag{7}$$

Using the assumption of nondivergent flow, a transport function ψ^* is introduced

$$M_y = \frac{\partial \psi^*}{\partial x} , \quad M_x = - \frac{\partial \psi^*}{\partial y} ,$$

and cross differentiation leads to

$$\frac{\partial}{\partial t} \nabla^2 \psi^* + \beta \frac{\partial \psi^*}{\partial x} + k \nabla^2 \psi^* = - \frac{\partial \tau_x}{\partial y} + \frac{\partial \tau_y}{\partial x} \tag{8}$$

where $\beta = \dfrac{\partial f}{\partial y}$.

Using the wind stress distributions for τ_x and τ_y given above, equation (8) becomes

$$\frac{\partial}{\partial t} \nabla^2 \psi^* + \beta \frac{\partial \psi^*}{\partial x} + k \nabla^2 \psi^* = \left(- \frac{\tau_{x_0} \pi}{L} \sin \frac{\pi y}{L} - \frac{\Delta \tau_{y_0}}{B} \right) e^{i \omega t} . \tag{9}$$

In order to find a simple analytical solution of the problem, the assumption is made that the surface circulation in the northern part of the Indian Ocean is strictly periodical. This assumption, which has been used previously (Phillips, 1966), seems to be well justified for a first-order approach to the problem. Hence, one can use the separation

$$\psi^*(x, y, t) = \psi(x, y) e^{i \omega t} . \tag{10}$$

Physically this means that the monsoon currents are forced oscillations having the same frequency ω as the driving force, which is the wind.

This separation leads to the equation

$$(k + i\omega) \nabla^2 \psi + \beta \frac{\partial \psi}{\partial x} = - \frac{\tau_{x_0} \pi}{L} \sin \frac{\pi y}{L} - \frac{\Delta \tau_{y_0}}{B} , \tag{11}$$

which is linear and can therefore be split into two equations:

$$\nabla^2 \psi_z + \frac{\beta}{k + i\omega} \frac{\partial \psi_z}{\partial x} = - \frac{\tau_{x_0} \pi}{L(k + i\omega)} \sin \frac{\pi y}{L} , \tag{11a}$$

which describes the effect of the *zonal* wind-stress component; and

$$\nabla^2 \psi_m + \frac{\beta}{k + i\omega} \frac{\partial \psi_m}{\partial x} = - \frac{\Delta \tau_{y_0}}{B(k + i\omega)} , \tag{11b}$$

which describes the effect of the *meridional* wind-stress component.

The analytical solutions of (11a) and (11b) can then be superimposed in order to get a complete solution of equation (9). For both equations the boundary conditions are:

$$\psi(0, y) = \psi(B, x) = \psi(x, 0) = \psi(x, L) = 0 \, . \tag{12}$$

A. Solution of the Zonal Problem

Equation (11a) is essentially the same as the one solved by Stommel (1948), except that the constants in our case are complex. Its general solution is

$$\psi_z = \Sigma(XY) + \frac{\tau_{x_0}L}{(k + i\omega)\pi} \sin \frac{\pi y}{L} \tag{13}$$

where

$$X = p_j e^{A_{1j}x} + q_j e^{A_{2j}x} \tag{14a}$$

$$Y = c_j \sin n_j y + d_j \cos n_j y$$

$$A_{1,\,2j} = -\frac{\beta}{2(k + i\omega)} \pm \sqrt{\left(\frac{\beta}{2(k + i\omega)}\right)^2 + n_j^2} \, , \tag{14b}$$

and c_j, d_j, p_j, and q_j are undetermined constants.

This very general solution reduces to a simple closed form when the boundary conditions (12) are applied and the solution is given by

$$\psi_z(x, y) = -\frac{\tau_{x_0}L}{\pi(k + i\omega)} \sin \frac{\pi y}{L} (pe^{A_1 x} + qe^{A_2 x} - 1) \tag{15}$$

where

$$p = \frac{1 - e^{A_2 B}}{e^{A_1 B} - e^{A_2 B}} \, ; \quad q = 1 - p \tag{16}$$

$$A_{1,\,2} = -\frac{\beta}{2(k + i\omega)} \pm \sqrt{\left(\frac{\beta}{2(k + i\omega)}\right)^2 + \frac{\pi^2}{L^2}} \, . \tag{17}$$

B. Solution of the Meridional Problem

Equation (11b) can be written as

$$\nabla^2 \psi_m + a \frac{\partial \psi_m}{\partial x} + b = 0 \tag{18}$$

where

$$a = \frac{\beta}{k + i\omega} \quad \text{and} \quad b = \frac{\Delta\tau_{y_0}}{B(k + i\omega)} \, . \tag{19}$$

By using the coordinate system presented in Figure 21, a solution of the form

$$\psi_m = \sum_{n=1}^{\infty} A_n(x) \cos \left(\frac{\pi y}{2L} (2n - 1)\right)$$

can be used to solve equation (18). This leads to

$$\sum_{n=1}^{\infty} \left\{ A_n'' - \frac{(2n - 1)^2 \pi^2}{4L^2} A_n + a A_n' \right\} \cos \left(\frac{\pi y}{2L} (2n - 1)\right) = -b \, . \tag{20}$$

By using the boundary conditions (12), it can be seen that

$$\psi_m(x, -L) = \psi_m(x, L) = 0$$

is already satisfied.

Furthermore, it is required that

$$\psi_m(0, y) = \sum_{n=1}^{\infty} A_n(0) \cos\left(\frac{\pi y}{2L}(2n - 1)\right) = 0 \qquad (21)$$

$$\psi_m(B, y) = \sum_{n=1}^{\infty} A_n(B) \cos\left(\frac{\pi y}{2L}(2n - 1)\right) = 0 , \qquad (22)$$

which leads to the boundary condition

$$A_n(0) = A_n(B) = 0 . \qquad (23)$$

In order to determine the coefficients A_n in (20), it is necessary that

$$\sum_{n=1}^{\infty} \alpha_n \cos\left(\frac{\pi y}{2L}(2n - 1)\right) = -b . \qquad (24)$$

The Fourier expansion (24) can be inverted, since it is noted that all cosine functions are harmonics of the interval $-2L < y < 2L$. Multiplying (24) by $\cos\left(\frac{\pi y}{2L}(2j - 1)\right)$ and integrating from $-2L$ to $+2L$ give

$$4L\alpha_j = -b\int_{-L}^{L} \cos\left(\frac{\pi y}{2L}(2j - 1)\right) dy + b\int_{L}^{3L} \cos\left(\frac{\pi y}{2L}(2j - 1)\right) dy . \qquad (25)$$

The integration leads to

$$\alpha_j = \frac{2b(-1)^j}{\pi(2j - 1)} . \qquad (26)$$

In order to obtain the solution of (20), the following equation has to be solved:

$$A_n'' + aA_n' - \frac{(2n - 1)^2\pi^2}{4L^2} A_n = \frac{4b(-1)^n}{\pi(2n - 1)} , \qquad (27)$$

taking into account the boundary conditions (23).

The solution of (27) is

$$A_n(x) = \frac{(-1)^n}{(2n - 1)^3} \frac{8bL^2}{\pi^3} \left[\frac{e^{\lambda_1 x}(e^{\lambda_2 B} - 1) + e^{\lambda_2 x}(1 - e^{\lambda_1 B})}{e^{\lambda_2 B} - e^{\lambda_1 B}} - 1\right] . \qquad (28)$$

Hence, the solution of the equation for the meridional problem is

$$\psi_m(x, y) = \sum_{n=1}^{\infty} \frac{(-1)^n}{(2n - 1)^3} \frac{8bL^2}{\pi^3} \left[\frac{e^{\lambda_1 x}(e^{\lambda_2 B} - 1) + e^{\lambda_2 x}(1 - e^{\lambda_1 B})}{e^{\lambda_2 B} - e^{\lambda_1 B}} - 1\right] \cos\left(\frac{\pi y}{2L}(2n - 1)\right) , \qquad (29)$$

where

$$\lambda_{1, 2} = -\frac{a}{2} \pm \sqrt{\frac{a^2}{4} + c} ,$$

$$a = \frac{\beta}{k + i\omega} , \quad b = \frac{\Delta\tau_{y_0}}{B(k + i\omega)} , \quad c = \frac{(2n - 1)^2\pi^2}{4L^2} . \qquad (30)$$

The solution represents an alternating series that, owing to the factor $\frac{1}{(2n - 1)^3}$, converges rapidly.

III. DISCUSSION OF THE ZONAL SOLUTION

In order to discuss the time-dependent behavior of the circulation and the effects of friction and the rotation of the earth, we must take into consideration the complete solution

$$\psi_z^*(x, y, t) = \psi_z(x, y)e^{i\omega t} \tag{31}$$

where $\psi_z(x, y)$ is given by expression (15).

A. Influence of Dissipation and Coriolis Force

Solution (31) was evaluated on an IBM 360 computer. A standard program to compute complex numbers was used. In order to discuss the influence of dissipation and Coriolis force in an effective way, the distribution of the mass transport is given in an east-west direction along the latitude circle $y_0 = \dfrac{L}{2}$. Thus

$$\psi_z^*(x, t) = -\frac{\tau_{x_0}L}{(k + i\omega)\pi}(pe^{A_1 x} + qe^{A_2 x} - 1)e^{i\omega t}, \tag{32}$$

where the complex quantities p, q, A_1, and A_2 correspond to those given in (16) and (17).

This quantity $\psi_z^*(x, t)$ can be interpreted as the total meridional transport (positive sign for clockwise circulation, negative sign for counterclockwise circulation) between $x = 0$ and $x = B$. The quantity $\psi_z^*(x, t)$ at a certain fixed time corresponds to the function $X(x)$ given by Munk (1950, Fig. 3). Each figure (Figs. 22–28) shows three sets, with each set consisting of 12 curves; each curve represents the volume transport for one month. If β is considered to be a variable parameter and the viscous parameter k is kept constant (Fig. 22), one notes that the number of peaks of the envelopes of all 12 curves decreases from 12 to 10 when β increases from 2.26×10^{-13} to 2.29×10^{-13} cm^{-1} sec^{-1}.

For an additional demonstration of the varying number of peaks that have varying values of β, three cases where $\beta > 2.29 \times 10^{-13}$ have been selected (*see* Table 3 *and* Fig. 28). These cases are unrealistic since β can never attain a value greater than 2.29×10^{-13} cm^{-1} sec^{-1}. The selected values for β would then mean that the radius of the earth would be smaller than R. Nevertheless, this case demonstrates that the number of peaks is a function of β.

TABLE 3:   Variation of Peaks with Varying β

$\beta = 2.26 \times 10^{-13}$ cm^{-1} sec$^{-1} \cong 8°\ 40'$ latitude $\equiv$ 12 peaks	
$\beta = 2.28 \times 10^{-13}$ cm^{-1} sec$^{-1} \cong 4°\ 14'$ latitude $\equiv$ 11 peaks	see Fig. 22
$\beta = 2.29 \times 10^{-13}$ cm^{-1} sec$^{-1} \cong 0°$ latitude $\equiv$ 10 peaks	
$\beta = 2.45 \times 10^{-13}$ cm^{-1} sec$^{-1} \equiv$ 3 peaks	
$\beta = 2.46 \times 10^{-13}$ cm^{-1} sec$^{-1} \equiv$ 2 peaks	see Fig. 28
$\beta = 2.48 \times 10^{-13}$ cm^{-1} sec$^{-1} \equiv$ 1 peak	

These results indicate that the rotation of the earth, respectively the variation of the Coriolis parameter with latitude, acts as an externally imposed ordering mechanism. As Veronis (1966a) states, " . . . an ocean basin on a uniformly rotating plane system has no ordered response of the type which is characteristic of a basin on a β-plane or on a sphere." An additional effect of this mechanism is a slight difference among the amplitudes of the mass transport. Compare also with Figure 22. Another interesting feature of nonstationary currents is shown also in the sequence of Figures 22 to 28. Figure 28, in which the viscous parameter $k = 0$, corresponds to the "nonstationary Sverdrup-relation" (*compare* Welander [1959]) of the type

$$\frac{\partial}{\partial t} \nabla^2 \psi + \beta \frac{\partial \psi}{\partial x} = \operatorname{curl} \tau \, .$$

On the one hand, Figure 28 does not indicate any westward intensification in the absence of a dissipative term while Figures 22 to 27 show that a noticeable east-west asymmetry is achieved only by increasing the viscous parameter k. On the other hand, a series of analogous computations, with $\beta = 0$ and with increasing values for k, leads to the result that the envelopes for all cases show a symmetrical picture, with one peak at the center of the basin but without a westward intensification. The computations show furthermore that the transport rates decrease as k increases; and that, for $k = 1 \times 10^{-7}$ sec^{-1}, they are in the same order of magnitude as for the cases when $\beta \approx 2.2 \times 10^{-13}$ cm^{-1} sec^{-1}.

From these calculations, it is concluded that in a nonstationary circulation — or at least in an oscillating circulation as in the Indian Ocean — the rotation of the earth alone cannot produce a westward intensification. In order to produce an east-west asymmetry and the associated boundary currents, frictional phenomena must be involved. Another conclusion has been pointed out in a private communication by Professor R. S. Arthur. It is based on the result that periodic solutions for a closed basin can be obtained with $k = 0$. This would suggest that the frictional dissipation associated with westward intensification is not needed because the driving wind stress curl oscillates between positive and negative values in this model.

In order to reduce our parameters to those commonly used, we nondimensionalize the time-dependent equation for the zonal problem

$$\frac{\partial}{\partial t} \nabla^2 \psi_z^* + \beta \frac{\partial \psi_z^*}{\partial x} + k \nabla^2 \psi_z^* = \frac{\tau_{x_0} \pi}{L} \sin \frac{\pi y}{L} e^{i \omega t} \tag{33}$$

by substituting

$$x = l x'; \quad y = l y'; \quad t = t' \omega^{-1}; \quad \psi_z^* = \tau_{x_0} \pi \beta^{-1} \psi_z^{*'} \tag{34}$$

that leads to

$$Ro \, \frac{\partial}{\partial t'} \nabla^2 \psi_z^{*'} + \frac{\partial \psi_z^{*'}}{\partial x'} + \epsilon \nabla^2 \psi_z^{*'} = - \frac{l}{L} \sin \pi y' e^{i t'} \tag{35}$$

where

$$Ro = \frac{\omega l}{\beta l^2} = \frac{\omega}{\beta l}; \quad \epsilon = \frac{k}{\beta l} \, .$$

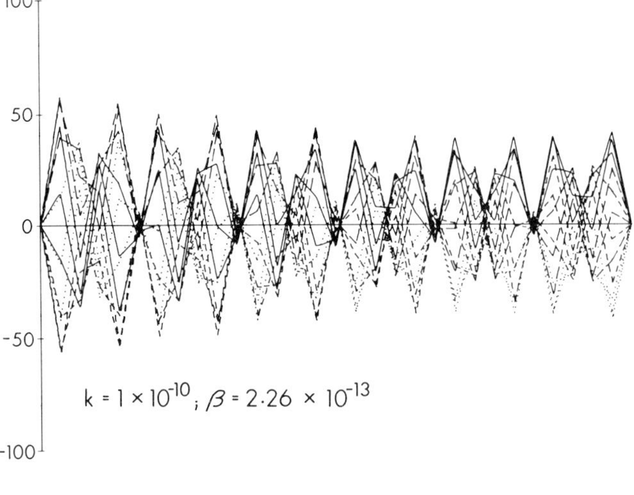

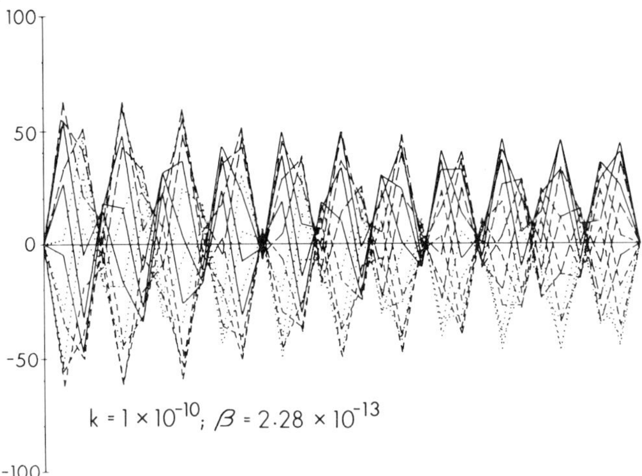

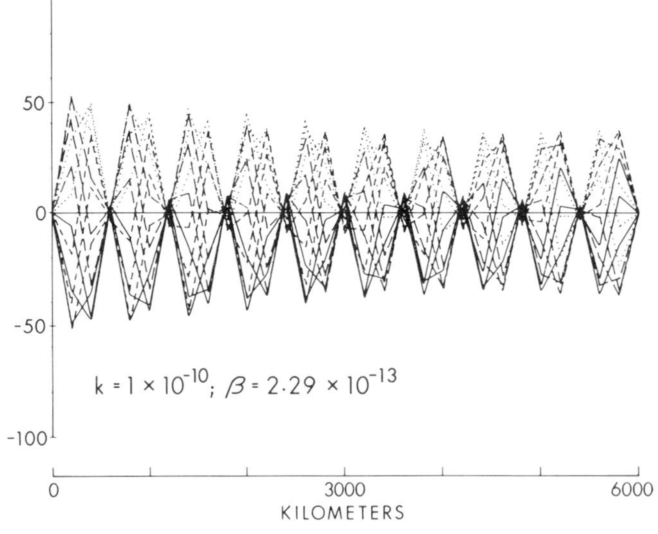

FIGURES 22–28: Distribution of the monthly mass transports in meridional direction along the latitude circle $y = (L/2)$. Positive quantities correspond to clockwise (Southwest Monsoon situation) and negative quantities correspond to counter-clockwise (Northeast Monsoon situation) sense of circulation in the model ocean. The ordinate gives volume transports in 10^6 m³ sec⁻¹, the full length of the abscissa corresponds to the length of the model ocean. Each figure shows three sets of 12 curves, one for each month. The following symbols were used for the four different *seasons:*

———————— February, March, April
—— —— —— May, June, July
. . . . August, September, October
— — — — — November, December, January

Maximum positive values (—— —— ——) correspond to July
Maximum negative values (— — — — —) correspond to January
Minimum values (. . . . .) correspond to October
Minimum values (————————) correspond to April

FIGURE 22

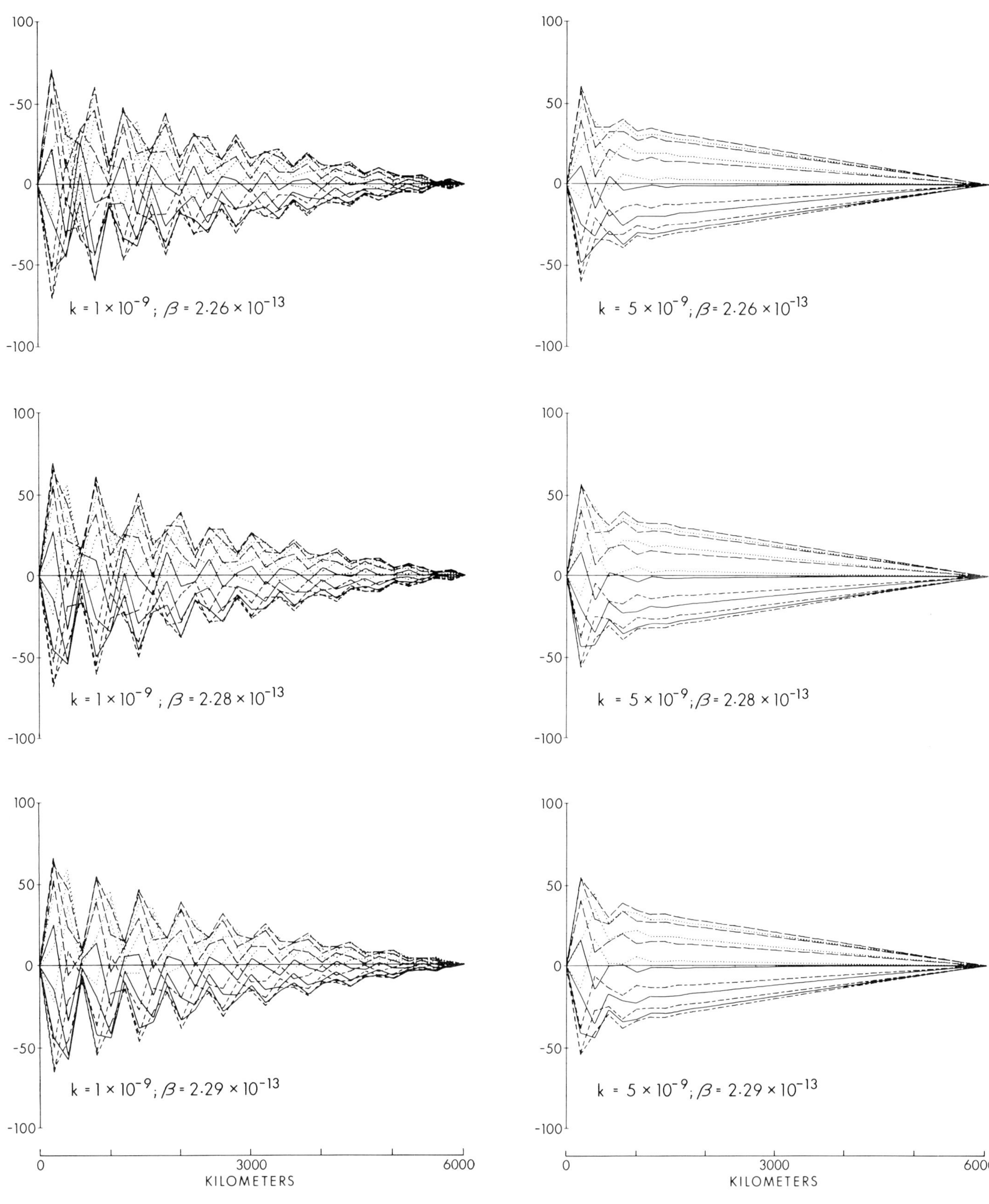

FIGURE 23

FIGURE 24

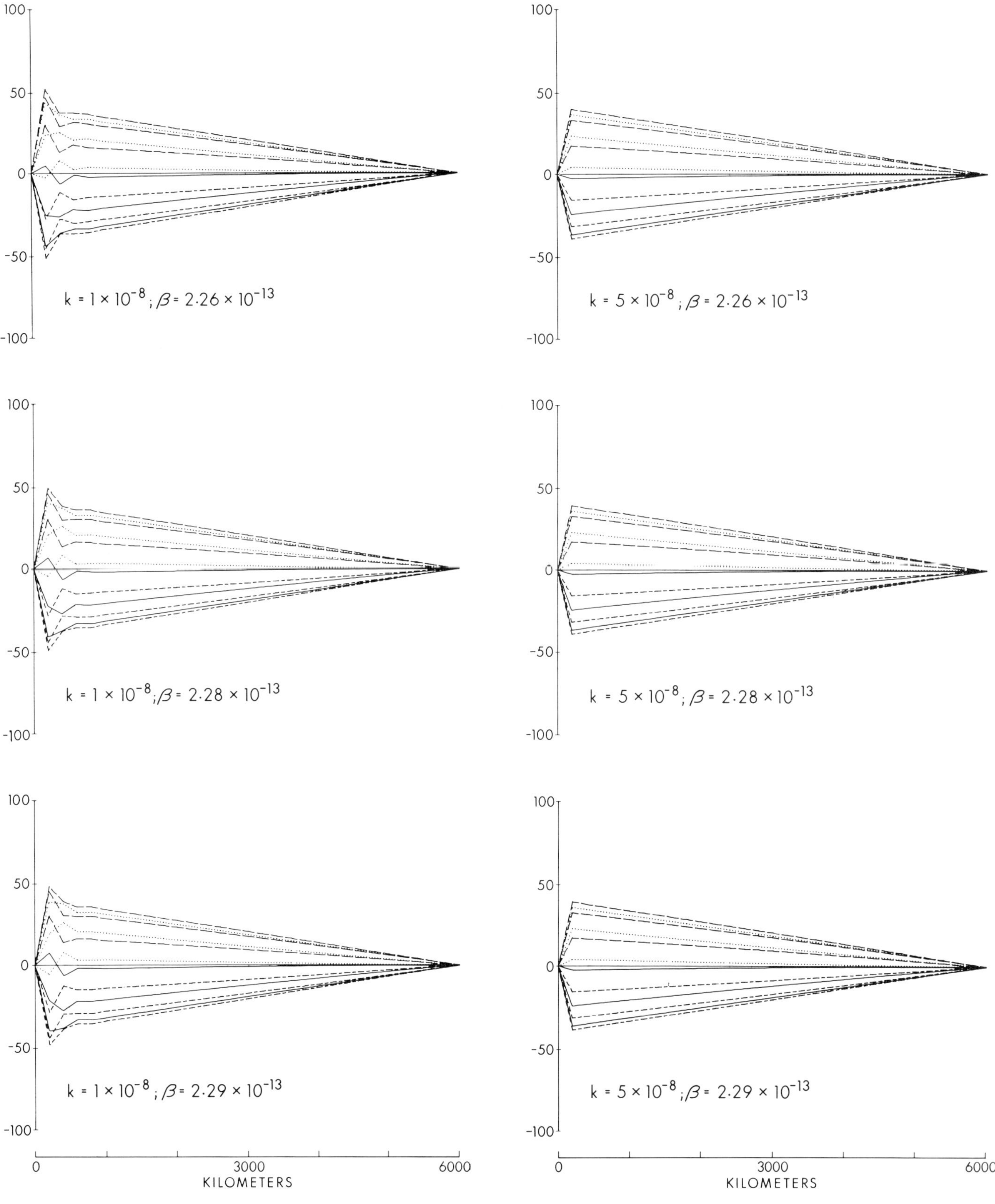

FIGURE 25   FIGURE 26

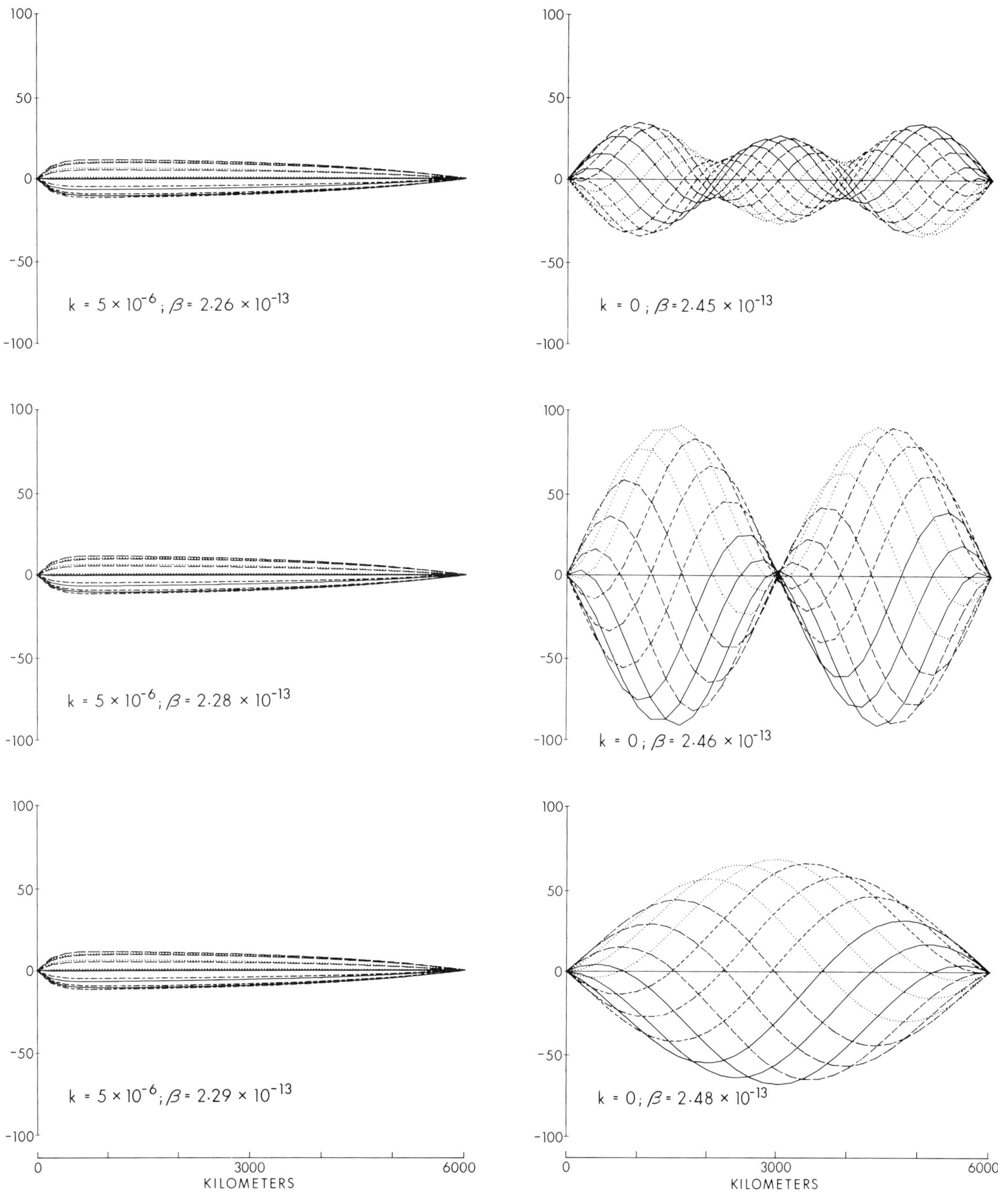

FIGURE 27

FIGURE 28

If the scale length l is comparable to the earth's radius, βl is equivalent to the Coriolis parameter. In this case, Ro may be interpreted as a Rossby number of the flow. ωl can be considered as a scale velocity V_s, which depends on the frequency of the driving force and the dimensions of the basin. In our case of a purely oscillatory behavior, however, $\dfrac{\omega}{\beta l}$ gives simply the relationship between the frequency of the driving force and the frequency of the earth's rotation, $Ro \simeq \dfrac{\omega}{f}$. The viscous parameter ϵ is of the same form as used by Veronis (1965, 1966a). Since we do not know what the value for k should be ("... or even if it is a meaningful parameter to use," Veronis [1965]), we shall discuss the influence of k on the width of the boundary region so that a reasonable k-value can be chosen. As Stommel (1965) states, the quantity $\epsilon = \dfrac{k}{\beta L}$ is simply $\dfrac{Ro}{Re}$. Thus, one can derive a relationship between the Reynolds number Re and the viscous parameter ϵ to be $Re = \dfrac{Ro\beta L}{k}$. Hence at fixed values for Ro, β, and L, Re is inversely proportional to k.

B. The Width of the Boundary Region

Solution (31) can be interpreted in terms of two superimposed planetary waves,

$$R_1(x, t) = -\frac{\tau_{x_0}L}{\pi(k + i\omega)} e^{i\omega t}(pe^{A_1 x} - 1) \tag{36}$$

and

$$R_2(x, t) = -\frac{\tau_{x_0}L}{\pi(k + i\omega)}(1 - p)e^{A_2 x}e^{i\omega t} \tag{37}$$

along a fixed latitude $y_0 = \dfrac{L}{2}$.

Figure 29 gives both waves for the parameters $k = 1 \times 10^{-8}$ and $\beta = 2.28 \times 10^{-13}$. This case corresponds to the one represented in Figure 25. If the amplitudes of both waves for each month are added, Figure 25 results. These waves can be interpreted as *forced* planetary waves with the frequency ω and are in contrast to the *free* planetary waves as described by Longuet-Higgins (1964), who uses the equation

$$\frac{\partial}{\partial t}(\nabla^2 \psi) + \beta \frac{\partial \psi}{\partial x} = 0$$

where the frequency is given by

$$\sigma = \frac{\beta k}{\kappa^2 + \eta^2}$$

and where κ and η are wave numbers in the x and y directions.

The portion of the R_2 wave where it is damped from its maximum to a value close to zero can be interpreted as the width of the boundary region.

In order to gain an insight into the functional dependence of this width from k and β, we have to split R_2 into the real and the imaginary parts.

By using the undetermined constants $p = p_1 + ip_2$ and $\dfrac{1 - p}{k + i\omega} = a_i + ia_2$, the imaginary part of $R_2(x, t)$ becomes

$$R_2(x, t) =$$

$$- \frac{\tau_{x0}L}{\pi} e^{d_2 x}\{a_2[\cos (e_2 x + \omega t) - p_1 \cos (e_2 x + \omega t) + p_2 \sin (e_2 x + \omega t)]$$

$$+ a_1[\sin (e_2 x + \omega t) - p_1 \sin (e_2 x + \omega t) - p_2 \cos (e_2 x + \omega t)]\} \qquad (38)$$

where

$$\left.\begin{aligned}
e_2 &= \frac{\omega\beta}{a} - \sqrt[4]{C_1^2 + C_2^2}\, \sin \phi_{1,\,2} \\[1em]
d_2 &= - \frac{\beta k}{a} - \sqrt[4]{C_1^2 + C_2^2}\, \cos \phi_{1,\,2} \\[1em]
a &= 2(k^2 + \omega^2) \\[1em]
c_1 &= \frac{\beta^2 L^2(k^2 - \omega^2) + \pi^2 a^2}{a^2 L^2} \\[1em]
c_2 &= \frac{2\omega k \beta^2}{a^2} \\[1em]
\phi_1 &= 3\pi + \alpha \quad \text{for} \quad k < \omega \\[0.5em]
\phi_2 &= 4\pi - \alpha \quad \text{for} \quad k > \omega \\[0.5em]
\alpha &= \text{ar} \tan \frac{|C_2|}{|C_1|}.
\end{aligned}\right\} \qquad (39)$$

For the R_1 wave, the analog expression to e_2 is given by

$$e_1 = \frac{\omega\beta}{a} + \sqrt[4]{C_1^2 + C_2^2}\, \sin \phi_{1,\,2}\,.$$

It can be shown that, even in the extreme case of $k = 0$, $e_{1,\,2} > 0$. Hence in all cases the R_1 and R_2 waves possess westerly phase velocities, which is in agreement with the results for free planetary waves as obtained by Longuet-Higgins (1964). The wave lengths of the $R_{1,\,2}$ waves is given by $\lambda_{1,\,2} = \dfrac{2\pi}{e_{1,\,2}}$. It is of interest to notice that in the frictionless case, where $k = 0$, a minimum wave length of $\lambda_{1,\,2} = \dfrac{4\pi\omega}{\beta}$ exists. For the values used above of $\omega = 2\pi(\text{year})^{-1}$ and β, a wave length of $\lambda_{1,\,2} \simeq 110$ km results.

An expression similar to (38) can be derived for R_1, but it is not given here. Ibbetson and Phillips (1967), in a model set up to explain experimentally produced Rossby waves, came to the conclusion that the two waves have their maximum amplitudes at opposite ends of the tank. In the present case, however, owing to the term -1 in (15) for both waves, the maximum amplitudes are found at the western boundary. It must be noted that these amplitudes have opposite signs.

If we consider the boundary region as the region where the maximum amplitude of the boundary current is damped off to $e^{-\pi} \simeq 1/23$ of its initial value, the width x_0 of this region is then given by

$$x_0 = - \frac{\pi}{d_2}\,. \qquad (40)$$

Figure 30 represents x_0 as a function of the viscous factor k, with β as a variable parameter. The set of parabolas on a semilogarithmic scale shows that the same width of the boundary region can be produced by two different values for the viscous parameter k. The extreme point of each curve is given at $k = \omega = 2 \times 10^{-7}$ sec^{-1}. For $k < 2 \times 10^{-7}$ sec^{-1}, the regime is of the inertial type. For $k > 2 \times 10^{-7}$ sec^{-1}, a viscous regime prevails.

With reference to the discussion above of the relationship between the Reynolds number and the viscous parameter k, Figure 30 shows that (i) for the *inertial type,* the width *increases* with the Reynolds number; and that (ii) for the viscous type, the width decreases with the Reynolds number. Result (ii) is in agreement with Bryan (1963), Fig. 3).

If one considers the amplitudes of the peaks in Figures 22–27, one recognizes that they steadily decrease as k increases. This feature is presented in a more concentrated form in Figure 31, where the maximum values of mass transport are given for different values of β and k. The convergence of the three lines indicates that the difference of the amplitudes — as caused by different values of β — is gradually wiped out by the increasing values of k.

If one draws a two-dimensional picture of the volume transport using the expression $\psi_z^*(x, y, t)$ instead of $\psi_z^*(x, t)$, one finds that each zero crossing corresponds to a circulation cell, where neighbor cells have either a cyclonic or an anticyclonic sense of rotation (Fig. 32). The occurrence of such cells, especially at the time of transition between both monsoon situations, is very similar to the western boundary vortices as discussed by Munk (1950) and Munk and Carrier (1950). Munk and Carrier say, " . . . whether the vortex will or will not be present depends only on the width of the ocean and the value of the eddy viscosity . . . ." By analogy, one can say that the existence and number of cells in the present case depend only on the value of the parameter k. According to Stommel (1965), the existence of the vortices is due to the use of the biharmonic operator $A\nabla^4\psi$ instead of $-k\nabla^2\psi$. The fact that the use of a simple dissipative law in the form of the term $-k\nabla^2\psi$ also leads to a vortex structure similar to the stationary model of Munk (1950) indicates that, in the present case, the vortex pattern is caused by the use of the acceleration term and not by the dissipative term. However, such a dissipative term, in one form or another, must be present when dealing with a barotropic model. This term represents the energy loss that is necessary to counteract the continued energy input through the wind stress. Even if the dissipation is represented through an oversimplified process, the principal physical effect is there; and, even if we err in detail, we have the necessary physical process incorporated.

In the light of recent publications, which include the nonlinear advective terms (Bryan [1963] and Veronis [1966b]) and which show an increasingly complex structure, with increasing Reynolds numbers, it seems that the hypothesized cells in our model would turn out to be even more complex if the nonlinear terms were included. In a recent publication on transient Rossby waves, Gates (1968) investigates the influence of the nonlinear terms on the transient circulation and shows that the meridional speed component has a cell structure similar to that discussed here. The nonlinear terms have the effect of inclining these cells slightly away from the NS direction toward an NNE-SSW direction. They also cause a slight NS-asymmetry that cannot be produced using a linear model.

Although Munk and Carrier made attempts to verify the existence of vortices from the observations, the reality of these vortices in quasi-steady circulation systems, as they occur in the Atlantic and the Pacific, is doubted among oceanographers (Stommel [1965]). In the Indian Ocean, however, conditions are completely different.

IV. THE MODEL CIRCULATION AND A COMPARISON WITH THE OBSERVATIONS

A comparison between the dynamic topographies and the present model for a homogeneous ocean is valid only up to a certain point. Owing to the lack of direct current observations, however, such comparisons have become customary in literature (Sverdrup [1947], Munk and Carrier [1950]). Certain implications of the validity of a barotropic model for the Indian Ocean are discussed at the end of this section.

The discussion in the previous part demonstrates the effect of dissipation on the mechanism of circulation. Let us postulate that the ultimate criteria for the choice of a reasonable numerical value of k are given by the existing oceanographic observations: (i) The gross features and the observed structure of the circulation cells as represented in the dynamic topographies should be taken into account. The observations indicate that a strongly oscillating circulation, as in Figure 22, does not exist in nature; they also show clearly that a single-gyre pattern, as in Figure 26, is not found either. (ii) The maximum volume transport during winter or summer should at least be approximated. Swallow and Bruce (1966) give a volume transport for the Somali Current of 50×10^6 m^3 sec^{-1}. (iii) The width of the Somali Current alone is given by Swallow and Bruce at 250 km. From *Monatskarten für den Indischen Ozean,* one finds, for both the summer and winter periods, a width for the general boundary region of 400 to 600 km. The last value includes the northwardly or southwardly directed countercurrents; hence, it is comparable to x_0 as defined in (40). The key diagrams for the choice of an appropriate value for k are given in Figure 30 and Figure 31.

According to Figure 30, a width of the boundary region of 500 km can be achieved either by $k = 10^{-8}$ sec^{-1} or by $k = 4 \times 10^{-6}$ sec^{-1}. Entering with these values into Figure 31, volume-transport rates of 50×10^6 m^3 sec^{-1}, respectively 12×10^6 m^3 sec^{-1}, result. Hence, the observed transport values can be achieved only by using the lower value for k. Furthermore, the choice of $k = 10^{-8}$ sec^{-1} leads to a circulation pattern containing several vortices (Fig. 25); while $k = 4 \times 10^{-6}$ sec^{-1} would lead to a single gyre-type circulation, as presented in Figure 27, with much too small transport rates. I am well aware that a choice of $k = 10^{-8}$ sec^{-1} deviates from values as they are usually given in the literature. Some examples are presented in Table 4.

Stommel's coefficient of friction, R, had to be divided by the depth, D, in order to be comparable. A numerical evaluation of Stommel's analytical result has been carried out: The dotted line in Figure 31 gives the maximum meridional volume transports along the center of the model basin as a function of k. The relatively small variation

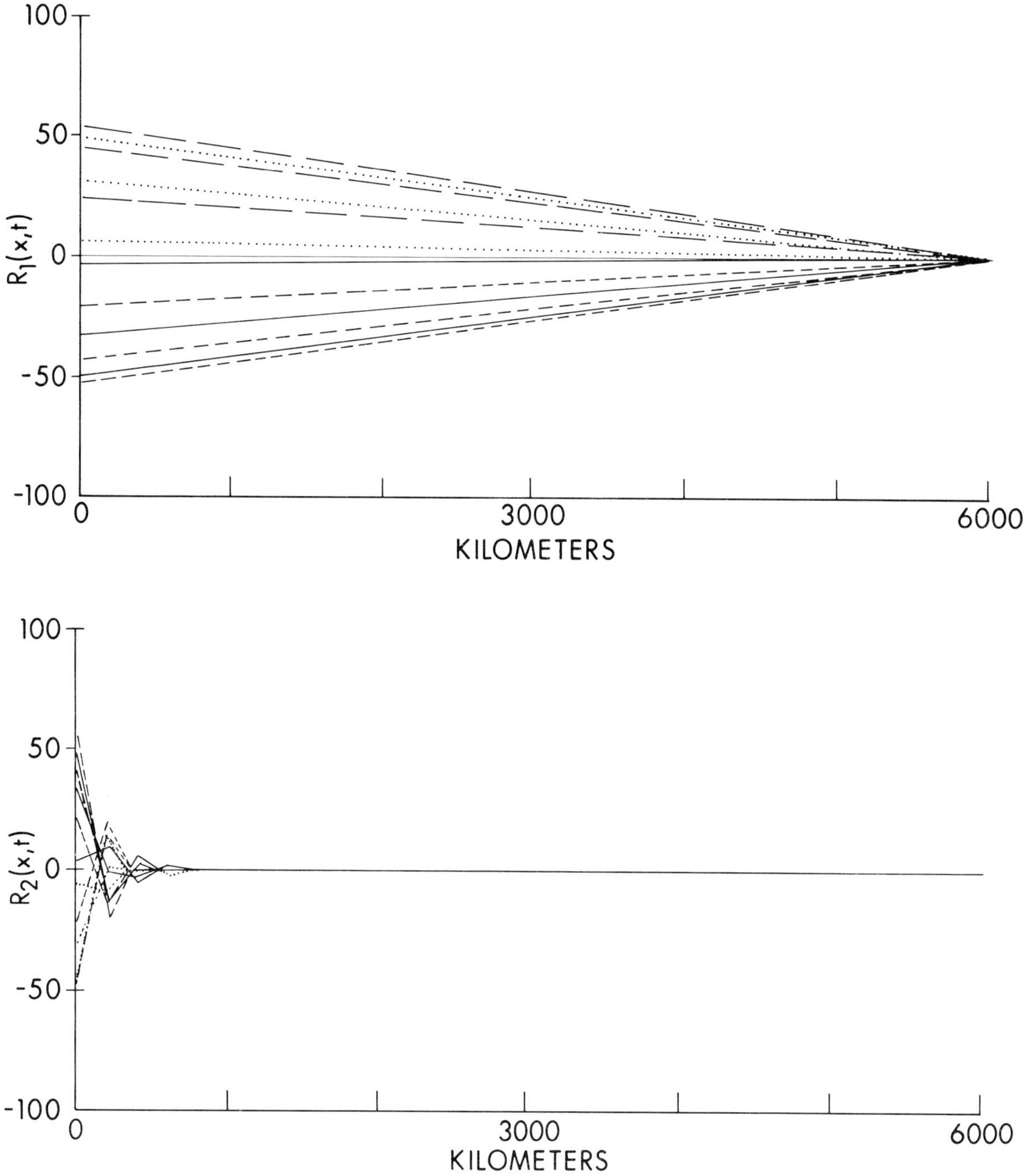

FIGURE 29: Interpretation of the case given in Figure 25 in terms of two planetary waves, long R_1 wave and short R_2 wave. Coordinate system and symbols same as given in Figures 22–28.

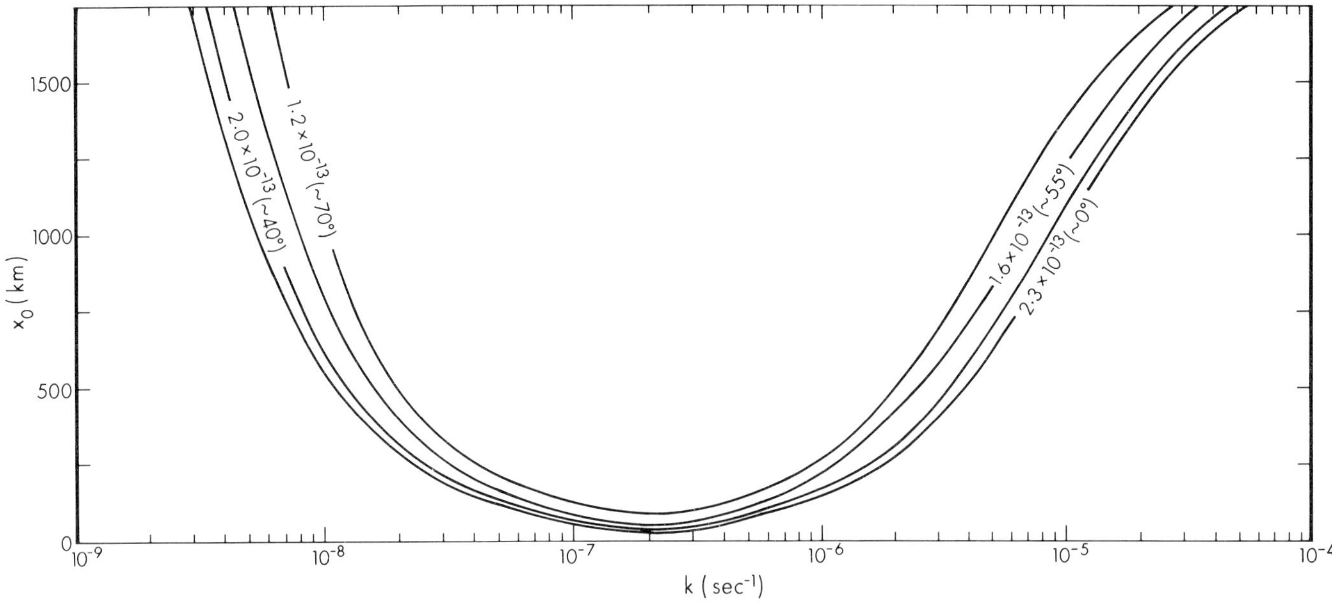

FIGURE 30: The width of the boundary region, x_0, as a function of the dissipative factor k, with B as a variable parameter.

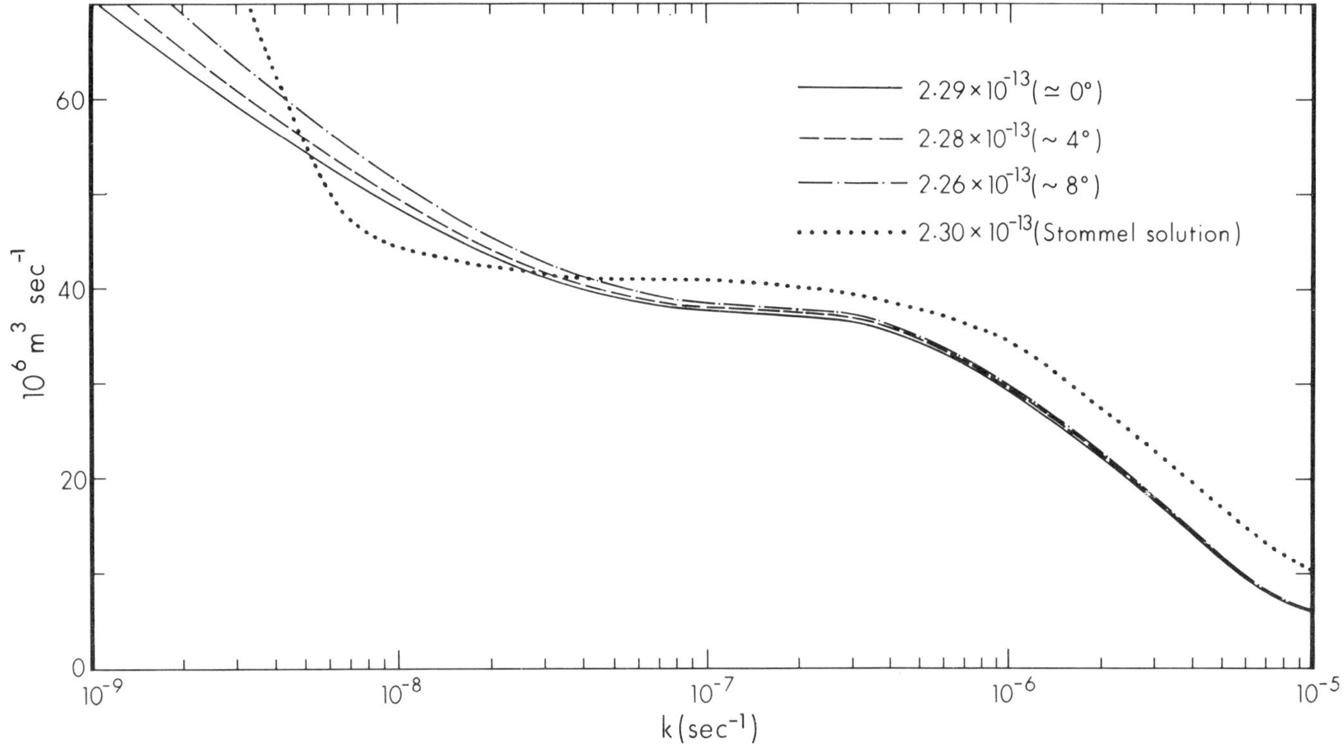

FIGURE 31: The maximum values for the meridional mass transport as a function of the dissipative factor k, with β as a variable parameter. For comparison .... gives the mass transport for a stationary model, based on the solution of Stommel (1948).

TABLE 4:   *k*-Values Given in the Literature

Author	*k*-Value
Stommel (1948)	1×10^{-6} sec^{-1}
Wyrtki (1960)	1×10^{-6} sec^{-1}
Wyrtki (1961)	2×10^{-6} sec^{-1}
Veronis (1965)	1×10^{-6} sec^{-1}
Veronis (1966)	0.8×10^{-6} to 2×10^{-6} sec^{-1}

over a wide range of k, from 10^{-6} to 10^{-8}, shows that, even in a stationary model, smaller k values yield reasonable results.

This means that dissipation and energy input by the wind are in balance. For values of $k < 1 \times 10^{-8}$ sec^{-1}, the transport values for the Stommel model shoot up to several 100×10^6 m^3 sec^{-1}, showing that this balance is disturbed and that the wind energy is not effectively dissipated. In the present model, the balance is influenced by the additional inertial term, which keeps the circulation from reaching exceedingly high values. As shown in Figure 28, transport rates do not reach 100×10^6 m^3 sec^{-1} even if no dissipation at all is involved.

The following discussion of the model circulation, especially with regard to the two-dimensional distributions in Figures 32 to 34, is based on the choice of $k = 10^{-8}$ sec^{-1}. The choice of a smaller value for k would have resulted in (i) more circulation cells, (ii) higher volume transports, and (iii) considerably higher phase lags between wind and mass transport. The circulation patterns are presented in monthly intervals, from July to January, in Figures 32 to 34. During the rest of the year, the pattern is repeated with opposite signs. The cycle begins at $t = 0$, which corresponds to April, according to (2a). To give a more detailed picture of the rapid changes during the transition time, 10-day intervals have been selected for October. The numbers represent volume transports in million m^3 sec^{-1}. Positive values indicate clockwise rotation (Southwest-Monsoon type), and negative values indicate counterclockwise rotation (Northeast-Monsoon type). The analytical result was printed out for a rather coarse grid size of Δx, $\Delta y = 200$ km.

The maximum volume transports corresponding to the zonal component of the wind stress amount to 50×10^6 m^3 sec^{-1} (Fig. 32). The monthly pictures reveal the fact that the summer circulation gradually decays owing to the fading of the driving force. The region of highest volume transport shifts from west to east: during July it is found 200 km from the western boundary, and during September–October it is found approximately 400 km offshore. After 172 days, the current near the western boundary, in the region from the coast to a distance 250 km offshore, vanishes completely. After 182 days, an eddy with the opposite sense of rotation develops in the nearshore area. Its transport is approximately the same as in the eddy remaining from the summer. Ten days later the new boundary eddy becomes markedly stronger, and a second eddy east of it begins to form. Between them we find a last, weak remnant of the summer pattern. Following this development through November and December, one sees that the large-scale oceanic gyre is built up of at least two eddies. The phase lag between wind stress and mass transport has been computed not to exceed 10 days in the boundary region (Fig. 35). A maximum phase lag of 9.7 days occurs 400 km off the western boundary in the center of the basin. A second intermediate maximum of 3.3 days is found 1000 km offshore. Both maxima coincide with the centers of the first and second eddy.

The evaluation of the solution for the meridional problem (29) resulted in maximum volume transports of 3×10^6 m^3 sec^{-1} (Fig. 33). This amounts to less than one-tenth of the maximum transport produced by the zonal stress component. This is explained, of course, by the difference of change in wind stress for both components. However, as it has been pointed out before, it seems possible that a linear decrease of the stress from west to east presents an underestimate of the real conditions. If an exponential decrease in the meridional wind component during the summer occurs, this could possibly account for the extremely high velocities found near the Somali coast during the Southwest Monsoon time. Swallow and Bruce (1966) report that surface velocities of 5 to 6 knots have been observed near the Somali coast and that velocities up to 7 knots have been found by the research vessel, *Discovery*. The main feature of the circulation corresponding to the meridional component of the stress (Fig. 33) is the development of a small clockwise vortex near the center of the western boundary region; this vortex is included between two elongated counterclockwise vortices to the north and south. From August on, the circulation pattern more and more resembles the pattern of circulation corresponding to the zonal stress component. Figure 33 shows that a maximum value of 3×10^6 m^3 sec^{-1} is reached after 162 days, whereas the wind maximum occurs at the 91st day. Thus, a phase lag of approximately 70 days results.

It seems surprising that a simple linear wind-stress distribution leads to a complicated circulation pattern that, during its initial state, is completely different from the pattern caused by a sinusoidal stress distribution. It is safe to assume that every type of stress distribution leads to a pertinent type of oceanic circulation. As a matter of fact, the difference between Stommel's (1948) single-gyre solution and Munk's (1950) multiple-gyre solution is due only to the choice of different distributions of the zonal stress. It can be concluded that, for a first-order approach to problems of wind-driven ocean circulation, simple analytical functions lead to basic results that do not differ much from results obtained by using more sophisticated stress distributions. A more realistic approach to the problem in general would be to replace the stress function characterized by a single wave number by the spectrum of the wind stress covering a whole range of wave numbers.

The superposition of the zonal and of the meridional solution represents the total solution of equation (9) and is presented in Figure 34. During summer and winter, when zonal winds are fully developed, the meridionally induced circulation is negligible; during

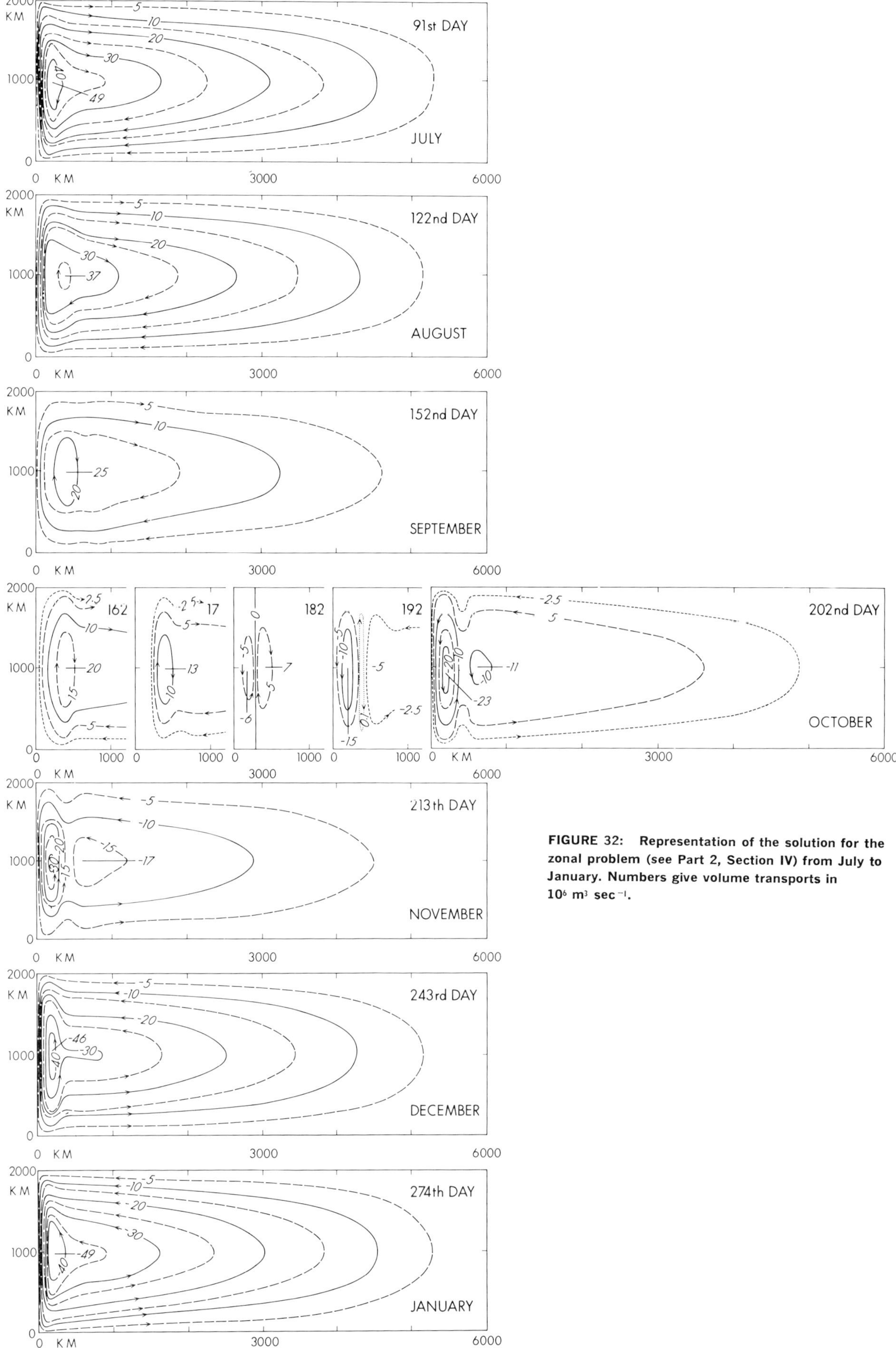

FIGURE 32: Representation of the solution for the zonal problem (see Part 2, Section IV) from July to January. Numbers give volume transports in 10^6 m³ sec⁻¹.

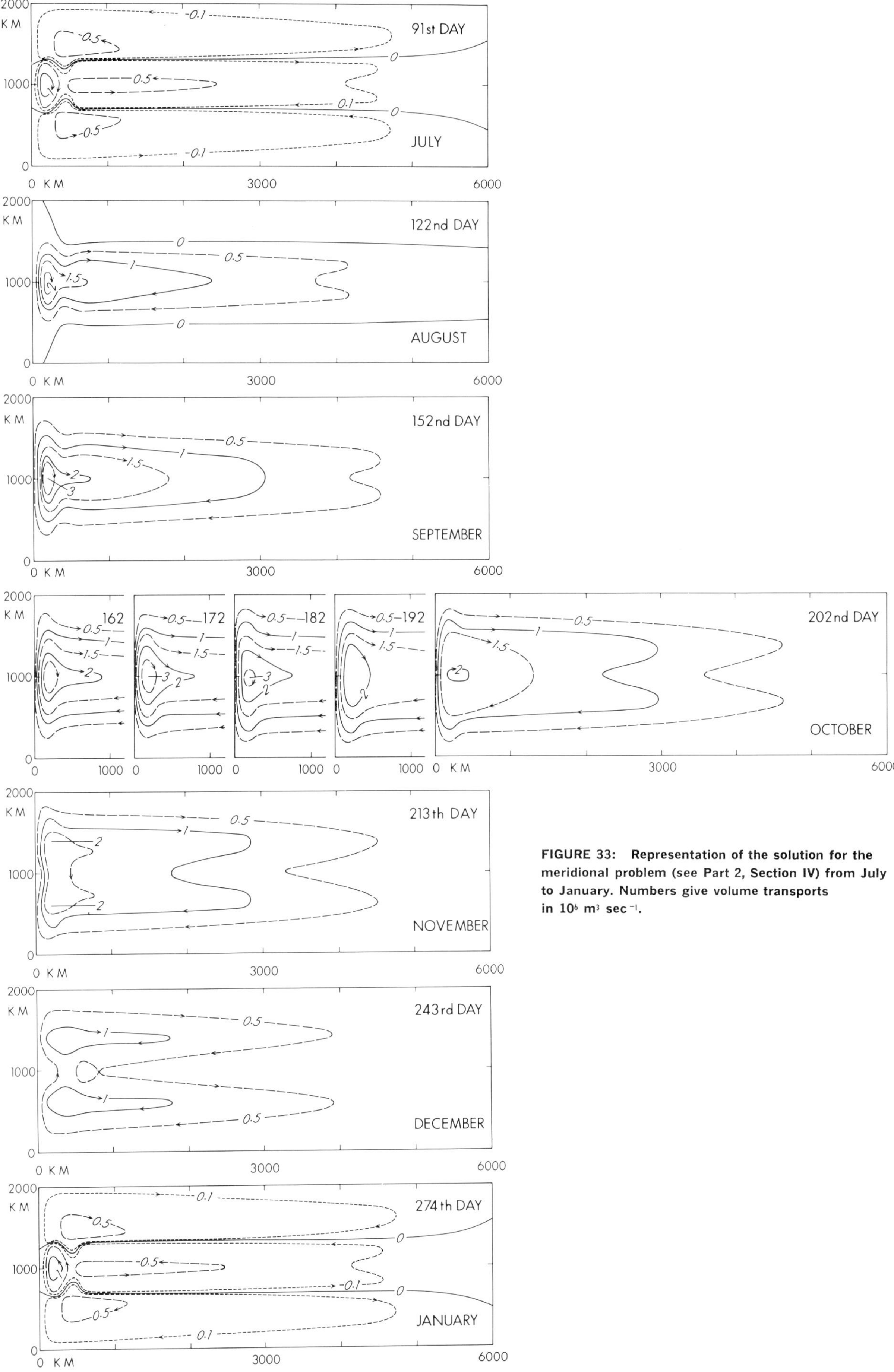

FIGURE 33:   Representation of the solution for the meridional problem (see Part 2, Section IV) from July to January. Numbers give volume transports in 10^6 m^3 sec^{-1}.

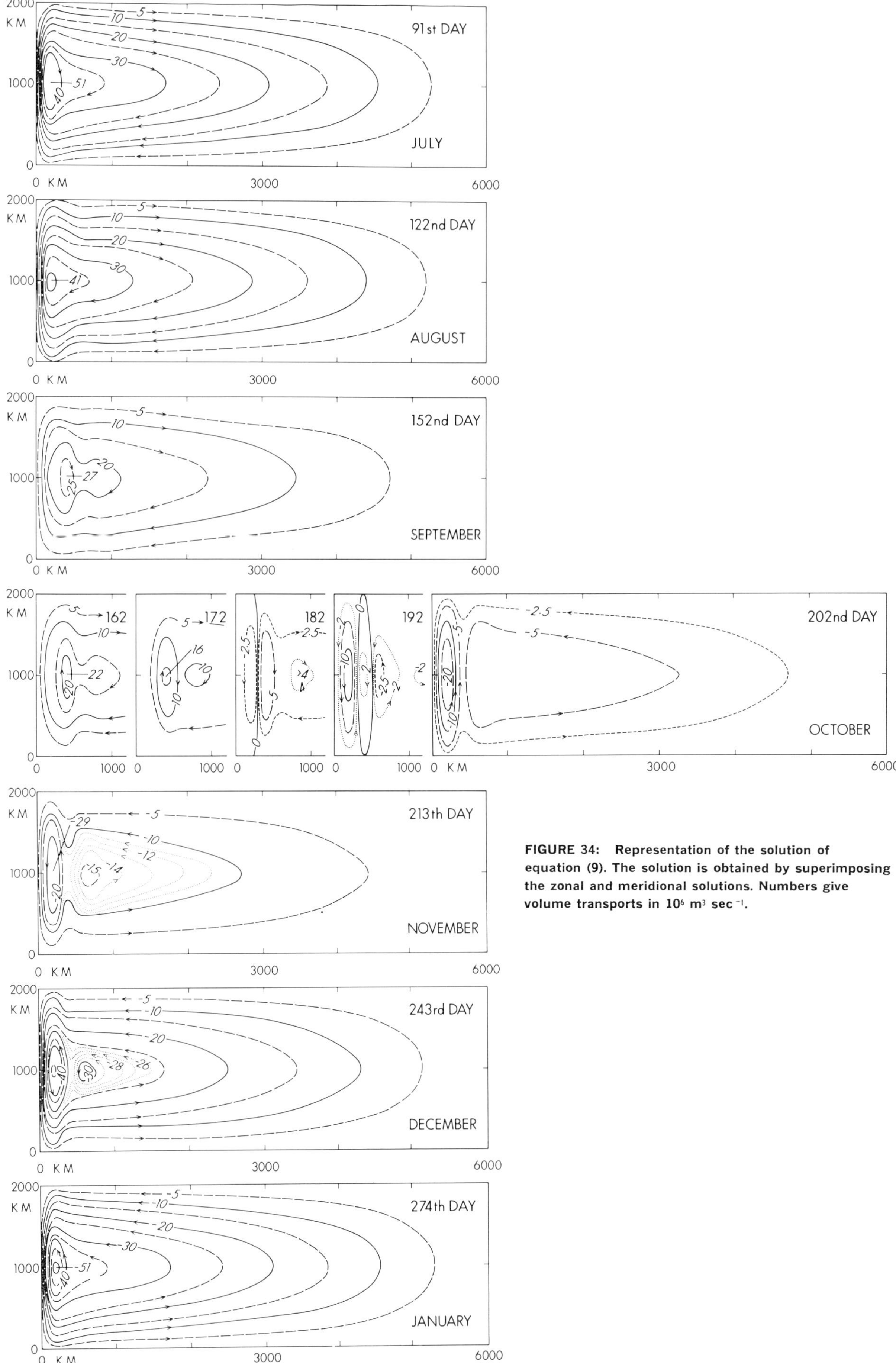

FIGURE 34: Representation of the solution of equation (9). The solution is obtained by superimposing the zonal and meridional solutions. Numbers give volume transports in 10^6 m^3 sec^{-1}.

the transition period, however, it has considerable influence. This is due to the phase lag of 70 days, so that relatively high values are reached when the zonally induced circulation reaches minimum values. As a consequence, the pattern during the transition becomes more complicated than in Figure 32, as represented in the pictures for the 182nd and 192nd days.

An additional feature shows up in Figure 34 (213th and 243rd days) : The vortices themselves show a crowding of the contour lines near the western edge, indicating that a westward intensification takes place in individual eddies remote from the western boundary.

A vertification of the theoretically derived vortex structure presents a major problem and cannot be done with a high degree of certainty. A useful link between theoretical investigations and observations in nature is provided by tank experiments. Ibbetson and Phillips (1967) conducted experiments on Rossby waves in a rotating annulus. Although their experiment differs in some important respects from the present model and from the real ocean, the results are in good qualitative agreement, especially when one compares Figure 5 in Ibbetson and Phillips with our Figure 34, October.

One surprising result of the present study is the rapid change in the circulation pattern and the rapid increase or decrease of the volume transport over short time intervals during the transition season. In the light of these results, it is clear that the present methods of oceanographic survey are insufficient to deal with such an extremely variable ocean. For better illustration, let us consider a hypothetical experiment. One might assume that the pattern given in Figure 34, for the 162nd to the 213th day, really occurs in nature. A modern research vessel like the *Atlantis II* is supposed to survey the western half of the basin during that time. Three east-west sections are carried out. The hydrographic stations are 100 km apart. To cover the distance of 3000 km, the vessel needs 14 days for each section, the whole survey lasting at least 42 days. During this time, the circulation pattern changes from the clockwise Southwest Monsoon circulation, going through a complex pattern in October, to the counterclockwise circulation during early November. One can easily see how confused a dynamic topography would look were it based on such a cruise, not to mention the additional confusion owing to the interference of internal waves and other sources of errors.

At the same time, the appearance of cell-like structures in the dynamic topographies during individual surveys as well as in average maps strongly supports the existence of vortices as a prominent feature of the circulation in the northern Indian Ocean.

It seems that even direct current measurements, sparse as they are, lead to similar conclusions. Martin *et al.* (1965) came to the following conclusions regarding their GEK measurements in the Indian Ocean:

Ainsi ce levé au G.E.K. s'accorde avec les grandes lignes bien connues de la circulation générale superficielle dans l'océan Indien; mais il décèle certains traits qui devront être l'objet de nouvelles études, tels que les trajets non strictement zonaux du "triplet" des courants et contre-courant équatoriaux.

The GEK measurements by and large confirm the structure of the surface currents in the Indian Ocean as they are generally known [from ship drift observations; author's note]. However, they also reveal certain patterns which must be studied more closely, such as the not strictly zonal flow of the triplet of equatorial currents and counter current. (Author's translation.)

These nonzonal components would indeed be explained by the vortex structure of the circulation.

The fact remains that the dynamic topographies show a great number of highs and lows of sea level, indicating cyclonic and anticyclonic vortices with orders of magnitude from several hundred to one thousand kilometers. None of the five charts shows that the northern part of the Indian Ocean during one particular season develops a single ocean-wide gyre as similar charts show for the Pacific Ocean or for the Atlantic Ocean.

This suggests that the choice of the factor $k = 10^{-8}\ \text{sec}^{-1}$ was perhaps too modest. It seems very well possible that the choice of a smaller k would have been adequate, perhaps, 5×10^{-9} (Fig. 24). As a consequence of such a choice, higher volume transports of approximately $60 \times 10^6\ \text{m}^3\ \text{sec}^{-1}$ for the boundary currents and a maximum phase lag of 20 days for the eddy near the boundary would result. This view is supported by the chart contoured for August–September 1963 (Fig. 14) and by the two sections for the volume transport computed by Bruce (Fig. 16). The surprisingly high transport rates as given by Bruce and the changing sense of rotation of the vortices in Figure 14 can theoretically be produced in two ways: (i) by using an extremely small value for the viscous parameter, such as $k = 10^{-9}\ \text{sec}^{-1}$, as in Figure 23; or (ii) by an eastward extension of the type of circulation given in Figure 34, 192nd day. In this case, cells with opposite senses of rotation are caused by the superimposed effect of the meridional stress component. It can be assumed, for example, that an exponential decrease of the stress from west to east, possibly in conjunction with a k slightly smaller than 10^{-8}, would simulate the observed effect not only during the transition period but also during summer.

One point that has not been discussed thus far deserves to be mentioned. It is the question of whether the monsoonal part of the Indian Ocean shows a barotropic or a baroclinic response to the stress exerted by the monsoon winds. Again, it does not seem to be possible to give an answer based on the available oceanographic observations.

From an earlier investigation by Veronis and Stommel (1956), we know, however, that wind changes of periods much longer than a year result in a baroclinic response of the ocean, in which case the present barotropic model would not be applicable.

This leads to the question of what other periods besides the basic period of one year are involved. It is a fact that the Southwest Monsoon season lasts longer than the Northeast Monsoon season and that higher wind speeds occur during summer than during winter (Fig. 6). This means that the monsoons not only contain the basic period of one year but also include a number of higher harmonics with periods of less than one year. According to Veronis and Stommel, one must then expect a barotropic response. It seems, therefore, that the present model is partially applicable to that ocean.

In order to test the results of the present investigation and to make progress in the understanding of transient processes, future theoretical studies should include two steps: (i) an investigation of a two-layer model that must include dissipation and lateral boundaries to find out how a quasi-baroclinic model responds to wind periods of one year, especially to determine whether a similar vortex structure results; (ii) an investigation of the effect of a combination of different wind periods. These could be represented in terms of a truncated Fourier series, or, preferably, by using a characteristic spectral distribution for the wind stress. A step in this direction was made in several recent Russian investigations (Belyaev, 1967), where the dependence of the spectra of the flow-velocity components on the spectrum of the tangential wind force was determined.

V. SOME CONSIDERATIONS ON KINETIC ENERGY

In order to get an approximate idea of the theoretical distribution of the kinetic energy for the monsoonally induced circulation (only for zonal problem), somewhat arbitrary assumptions about the vertical distribution of density and current velocity must be made. For the density, a simple linear increase with depth was assumed:

$$\rho(z) = \rho_0 - \frac{\Delta\rho}{h} z \tag{41}$$

where $\Delta\rho = \rho_h - \rho_0$ with $\rho_h = 1.027$ g cm^{-3}

$$\rho_0 = 1.023 \text{ g cm}^{-3}$$

$$h \simeq -500 \text{ m} .$$

For the horizontal veloicty components, simple exponential forms were assumed:

$$u(z) = u_0 e^{\frac{\pi}{h} z}; \quad v(z) = v_0 e^{\frac{\pi}{h} z} \tag{42}$$

where z is directed from the sea surface upward.

According to equation (5), the horizontal transports are given by

$$M_x = -\psi_y^* = -\int_{-h}^{0} \rho(z)u(z)\, dz$$

$$M_y = \psi_x^* = \int_{-h}^{0} \rho(z)v(z)\, dz .$$

Using assumptions (41) and (42), the surface velocities are given by

$$u_0 = Re\left\{\frac{-\psi_y^*}{H}\right\} \quad \text{and} \quad v_0 = Re\left\{\frac{\psi_x^*}{H}\right\} \tag{43}$$

where ψ_x^* and ψ_y^* are the derivatives from expression (31) with respect to x and y, and

$$H = \frac{\rho_0 h}{\pi} - \frac{\rho_0 h}{\pi} e^{-\pi} + \frac{h\Delta\rho}{\pi^2} + \frac{h\Delta\rho}{\pi}(1 + \pi)e^{-\pi} . \tag{44}$$

An inspection of (44) indicates that the last three terms can be neglected, thus indicating that the vertical density structure is relatively unimportant if compared to the vertical current profile.

The vertically averaged speeds for the upper layer are given by

$$\bar{u}(x, y, t) \; = \; \frac{u_0}{\pi}\,(1 - e^{-\pi}); \quad \bar{v}(x, y, t) \; = \; \frac{v_0}{\pi}\,(1 - e^{-\pi}) \tag{45}$$

and the kinetic energy per unit area, by

$$E(x, y, t) \; = \; \frac{1}{2}\,\Delta L\,\Delta B h \bar{p}[\bar{u}^2(x, y, t) + \bar{v}^2(x, y, t)]. \tag{46}$$

Taking into consideration (43) and (44), one recognizes that the kinetic energy is inversely proportional to h. If h decreases, the kinetic energy must increase, because the available wind energy is transferred into a shallower layer of water. The computed absolute figures for the kinetic energy are of a limited value owing to the rather arbitrary choice of h. However, it is of interest to consider the relative distribution of kinetic energy in the model basin.

Figure 36 shows same energy profiles for the month of May in the north-south direction. The profiles are distributed from the western to the eastern edge of the basin.

The absolute maximum of the kinetic energy is found at $y = \dfrac{L}{2}$ at the western boundary. This profile sharply decreases along the zonal boundaries. In the central part, from 1000 to 5000 km, the energy increases towards the northern and southern boundary. At $y = \dfrac{L}{2}$, a constant level of energy is found east of $x = 1000$ km.

Figure 37 gives the distribution of kinetic energy in the zonal direction for $y = \dfrac{L}{2}$ for the model months, May, June, and July. The main feature is the sharp drop in the energy levels from the western boundary region to the rest of the ocean. The values vary by a factor of the order of the magnitude of 10^5. A grid size Δx, $\Delta y = 50$ km, instead of 200 km, was used. As a result, one gains a more detailed insight into the energy distribution in the individual vortices. For each monthly curve, five intermediate maxima can be detected. In analogy to the westward crowding of the streamlines in each individual vortex, one finds that the peaks for the energy are situated asymmetrically to the western side. Furthermore, an eastward shift of the peaks in the individual cells can be seen from month to month. In the boundary region there exist considerable phase lags between the induced energy and the driving force. In the rest of the ocean the variation of the kinetic energy is nearly in phase with the wind.

The concentration of energy in the western region reflects the existence of circulation cells as discussed in the previous sections. From Figure 37 it is noted that the distances between the energy peaks amount to approximately 200 to 300 km. It is interesting that the investigation of Yampol'skii (1966) on the dependence of the velocity spectrum of a drift current on the spectrum of wind stress leads to similar results. Yampol'skii arrives at an expression which determines the spatial scales for which a transfer function ensures maximum transmission of wind energy to currents. Using appropriate values for the conditions in the Indian Ocean, one finds for an exchange coefficient, $\nu = 1 \times 10^8$ cm^2 sec^{-1}; a spatial scale of $L = 200$ km; and for $\nu = 5 \times 10^8$, $L = 450$ km. Hence, two investigations from two different points of view result in the maximum transmission of wind energy to currents for wind periods of one year takes place in circulation structures of a few hundred kilometers in size.

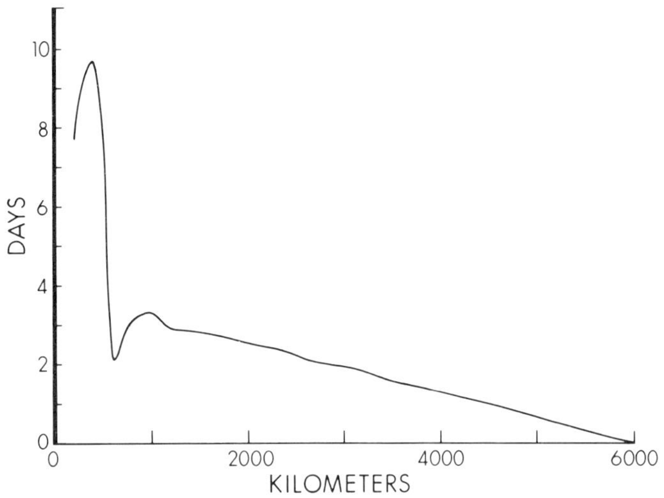

FIGURE 35:   The phase lag between the wind stress and the volume transport for the zonal solution. $k = 1 \times 10^{-8}$ sec^{-1}; $\beta = 2.28 \times 10^{-13}$ cm^{-1} sec^{-1}.

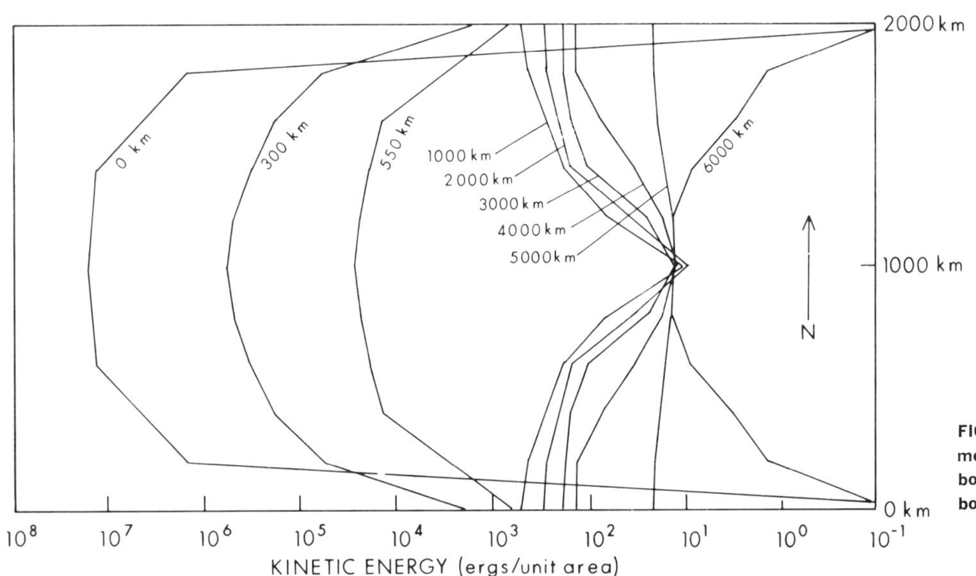

FIGURE 36:   Energy profiles in meridional direction from the western boundary (0 km) to the eastern boundary (6000 km) for May.

FIGURE 37:   Energy profiles in zonal direction along $y = (L/2)$ for May, June, and July.

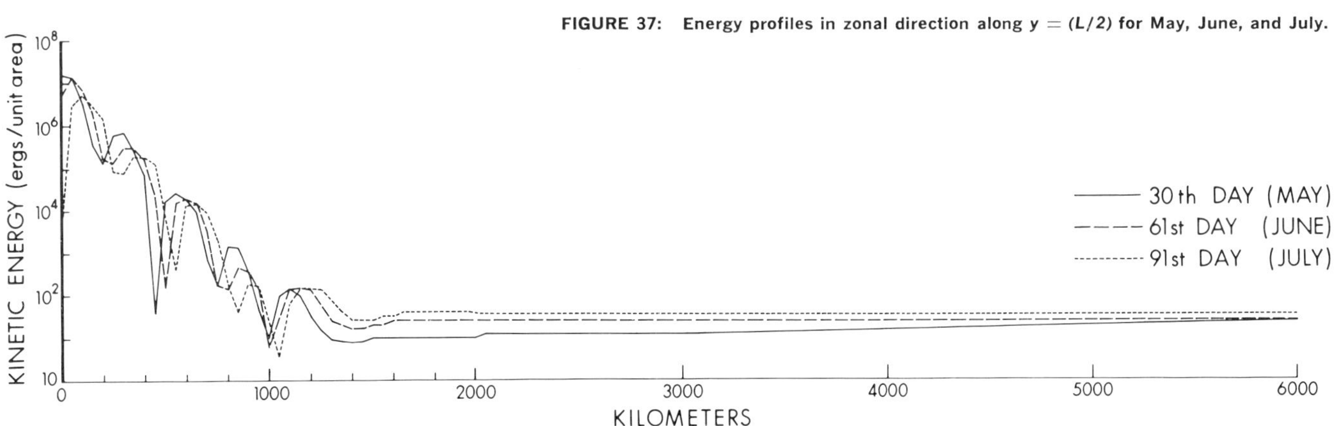

PART 3: *Summary and Conclusions*

1. This investigation is limited to the monsoonal part of the Indian
   Ocean and its seasonally changing current system. In order to estab-
   lish the region under consideration, the following definitions are
   given: (i) The extension of the monsoon winds and of the monsoonally affected surface currents is essentially limited to the area north
   of 10° S. (ii) The monsoonal circulation penetrates to a maximum
   depth of approximately 400 m in the western part of the ocean and
   shows minimum values of less than 100 m in the central part . (iii)
   Five characteristic time periods of the year have been defined during
   which quasi-consistent conditions prevail.

2. It was necessary to apply statistical criteria in order to remove errors
   and variations in the dynamic-height values. The dynamic topographies for the sea surface north of 20° S have been contoured for
   spring, early summer, late summer, fall, and winter. In addition to
   these average charts based on data from several years, one chart for
   the Arabian Sea has been contoured, using data from summer 1963
   only. The main features are: (i) The occurrence of a complex pattern of lows and highs common to all charts. A comparison with the
   sea-surface topographies of the Atlantic Ocean and the Pacific Ocean
   reveals that large gyres existing in these two oceans are not found
   in the northern part of the Indian Ocean and, vice versa, that the
   conditions prevailing in the Indian Ocean are found nowhere else.
   (ii) There is evidence in the average charts that the number of cyclonic or anticyclonic vortices is higher during the transition seasons
   than during the summer and winter season. (iii) A brief discussion
   of the dynamic topographies in terms of the conventionally known
   surface currents is given.

3. A theoretical model is presented in order to give a possible interpretation of the monsoonal circulation in the northern part of the
   Indian Ocean. The nonstationary, linear, barotropic model includes
   dissipation in the simple form $k\nabla^2\psi$; and, furthermore, it takes into
   account the time-dependent components of the zonal and the meridional wind stress. The zonal component of the wind stress is approximated by a cosine distribution, and the meridional component, by a
   linear distribution.

4. With the assumption that the monsoon currents are forced oscillations that have the same frequency as the driving wind stress, a

separation of the time variable allows one to find simple analytical solutions. The linearity of the remaining differential equation permits a split into two equations: one for the zonal and one for the meridional part of the problem. For both of these equations, analytical solutions — containing complex parameters — are given.

5. A detailed discussion of the zonal solution leads to the following conclusions: (i) The rotation of the earth acts as an externally imposed ordering mechanism causing a decreasing number of circulation cells while values of β increase. This effect can be recognized only if very small values of the frictional factor k are used. No westward intensification is produced under the absence of friction. (ii) East-west asymmetry and a decreasing number of circulation cells are achieved by increasing the values of k. (iii) The zonal solution can be interpreted as two superimposed forced Rossby waves that have the same frequency as the wind stress. Both waves possess westerly phase velocities. (iv) A discussion of the width of the western boundary region shows that for $k < \omega$ the regime is of the inertial type and for $k > \omega$ it is of the viscous type. The key diagram is given in Figure 30.

6. The governing parameter, k, has to be selected so that the pattern of the model circulation corroborates the observations that can be characterized by three main features: the cellular structure as presented in the dynamic topographies; the maximum mass transport of the boundary current during summer or winter; and the width of the western boundary region. A choice of $k = 1 \times 10^{-8}$ sec^{-1} (inertial type) leads to such results for the model that it is basically in agreement with the observations.

7. During July, the zonal solution yields maximum volume transports of 50×10^6 m^3 sec^{-1}. A maximum phase lag of 10 days occurs between wind and mass transport in the boundary region. With the fading wind, a gradual decay of the circulation occurs. The center of the gyre shifts from West to East during the decay process. A new eddy having the opposite sense of rotation begins to develop near the western boundary.

   The meridional solution yields maximum transport rates of 3×10^6 m^3 sec^{-1} during September. Hence there is a phase lag of approximately 70 days. A complex pattern of vortices develops in the northern and southern part of the basin. If fully developed, the pattern resembles the one found in the zonal solution.

   The meridional solution has a negligible contribution to the total solution during winter and summer when zonal winds are fully developed. During the transition period, however, it has considerable influence: the volume-transport rates are then increased, and the vortex structure becomes more complex. From October to December, respectively from April to June, there always exist several vortices arranged in an east-west direction in a way similar to that of the dynamic topographies.

8. The meridional volume transports as computed by Bruce (1968) for two *Atlantis II* sections show a changing pattern of northward and southward mass transport during August and September. This pat-

tern also finds its expression in the corresponding topography for summer 1963. It can be explained only if a value for $k < 10^{-8}$ sec^{-1} is used. The high rates of volume transport as computed by Bruce would then also be explained.

9.   The results of the model calculations indicate a complex and rapidly changing circulation pattern, which cannot be adequately surveyed through standard oceanographic techniques. Furthermore, the data available at present cannot answer the question of whether the monsoons cause a barotropic or baroclinic response of the ocean. Future investigations of the monsoonal circulation should include: (i) a two-layer model to simulate an ocean with a density stratification; (ii) a model taking into account a more realistic distribution of the wind stress in time and space; and (iii) especially designed cruises.

10.   Some considerations on kinetic energy are given, assuming simple vertical distributions for density and current. The highest values for the kinetic energy are found in the western part of the ocean, with each vortex showing an energy maximum shifting eastward from May to June. The central and eastern parts of the ocean show a constant low level of kinetic energy.

References

Belyaev, V. S., 1967: The dependence of the spectra of the velocity components of a wind-driven current on the spectrum of the tangential wind force. *Izv. Atmos. and Oceanic Phys.* 3(11):1217–1226, translated by J. Findlay.

Bruce, G. J., Jr., 1968: Comparison of zonal sections in the upper kilometer during the Southwest and Northeast Monsoons in the Arabian Sea. Unpublished results, by private communication.

Bryan, K., 1963: A numerical investigation of a nonlinear model of a wind-driven ocean. *J. Atmos. Sci.*, 20:594–606.

Defant, A., 1941: Die absolute Topographie des physikalischen Meeresniveaus und der Druckflächen, sowie der Wasserbewegungen im Atlantischen Ozean. *Deutsche Atlantische Expedition "Meteor" 1925–1927, Wissenschaftl. Erg.*, 6(2):191–260.

Defant, A., 1950: Reality and illusion in oceanographic surveys. *J. Mar. Res.*, 9:120–138.

Deutsches Hydrographisches Institut, 1960: *Monatskarten für den Indischen Ozean.* Publ. No. 2422, Hamburg.

Dietrich, G., and K. Kalle, 1957: *Allgemeine Meereskunde.* Eine Einführung in die Ozeanographie. Gebr. Borntraeger, Berlin-Nikolassee.

Düing, W., and W. D. Schwill, 1967: Ausbreitung und Vermischung des salzreichen Wassers aus dem Roten Meer und aus dem Persischen Golf. *"Meteor" Forsch. Erg.* A3:44–66.

Gates, W. L., 1968: A numerical study of transient Rossby waves in a wind-driven homogeneous ocean. *J. Atmos. Sci.*, 25(1):3–22.

Hydrographic Office, 1958: *Atlas of Surface Currents, Indian Ocean.* Publ. No. 566, Washington, D.C.

Ibbetson, A., and N. Phillips, 1967: Some laboratory experiments on Rossby waves in a rotating annulus. *Tellus*, 19(1):81–87.

Koninklijk Nederlands Meteorologisch Instituut, 1952: *Indische Oceaan. Oceanografische en Meteorologische Gegevens.* 2nd edition, Publ. No. 135, De Bilt.

La Violette, P. E., 1967: *Temperature, Salinity, and Density of the World's Seas: Bay of Bengal and Andaman Sea.* Naval Ocean. Off., Inf. Rept. No. 67-57, Washington, D.C.

Longuet-Higgins, M. S., 1964: Planetary waves on a rotating sphere. *Proc. Roy. Soc.* A279:446–473.

Martin, J.; P. Guibout; M. Crepon; and J. C. Lizeray: Circulation Superficielle dans L'Ocean Indien. Résultats de mesures faites a l'aide du courantometre a électrodes remorquées G.E.K. entre 1955 et 1963. *Cahiers Ocean.* XVII, Suppl. No. 3.

Munk, W., 1950: On the wind-driven ocean circulation. *J. Meteor.*, 7(2):79–93.

Munk, W. H., and G. F. Carrier, 1950: The wind-driven circulation in ocean basins of various shapes. *Tellus*, 2:158–167.

Phillips, N., 1966: Large-scale eddy motions in the Western Atlantic. *J. Geophys. Res.*, 71(16):3883–3891.

Ramage, C. S. (in press): Climate of the Indian Ocean. *In* H. Thomsen (ed.), *Climates of the Oceans, 15.* Elsevier Publ. Co., Amsterdam.

Reid, J. L., Jr., 1961: On the geostrophic flow at the surface of the Pacific Ocean with respect to the 1000-decibar surface. *Tellus*, 4:489–502.

Schmitz, H. P., 1964: Modellrechnungen zu winderzeugten Bewegungen in einem Meer mit Sprungschicht. *Deutsche Hydr. Zeit.*, **17**(5):201–232.

Schott, G., 1935: *Geographie des Indischen und Stillen Ozeans.* Verlag von C. Boysen, Hamburg.

Stommel, H. M., 1948: The westward intensification of wind-driven ocean currents. *Trans. Am. Geophys. Union*, **29**:202–206.

Stommel, H. M., 1965: *The Gulf Stream.* 2nd edition, University of California Press, Los Angeles.

Sverdrup, H. U., 1947: Wind-driven currents in a baroclinic ocean; with application to the equatorial currents of the eastern Pacific. *Proc. Nat. Acad. Sci.*, **33**(11):318–326.

Swallow, J. C., and J. G. Bruce, 1966: Current measurements off the Somali coast during the Southwest Monsoon of 1964. *Deep-Sea Res.*, **13**(5):361–888.

U. S. Department of Agriculture, 1938: *Atlas of Climatic Charts of the Oceans.* Weather Bureau Publ. No. 1247, Washington.

Veronis, G., 1963: An analysis of wind-driven ocean circulation with a limited number of Fourier components. *J. Atmos. Sci.*, **20**:577–593.

Veronis, G., 1965: On parametric values and types of representation in wind-driven ocean circulation studies. *Tellus*, **17**(1):57–84.

Veronis, G., 1966a: Generation of mean ocean circulation by fluctuating winds. *Tellus*, **17**(1):65–76.

Veronis, G., 1966b: Wind-driven ocean circulation — Part 1 and Part 2. *Deep-Sea Res.*, **13**(1):17–55.

Veronis, G., and H. Stommel, 1956: The action of variable wind stresses on a stratified ocean. *J. Mar. Res.*, **15**(1):43–75.

Warren, B. A., 1966: Medieval Arab references to the seasonally reversing currents of the North Indian Ocean. *Deep-Sea Res.*, **13**(2):167–171.

Welander, P., 1959: On the vertically integrated mass transport in the oceans. *In* B. Bolin (ed.), *Rossby Memorial Volume*, 95–101, New York and Oxford.

Wyrtki, K., 1960: The Antarctic Circumpolar Current and the Antarctic Polar Front. *Deutsche Hydr. Zeit.*, **13**(4):53–174.

Wyrtki, K., 1961: The thermohaline circulation in relation to the general circulation in the oceans. *Deep-Sea Res.*, **8**(1):39–64.

Yampol'skii, A. D., 1966: The dependence of the velocity spectrum of a drift current on the spectrum of the tangential wind stress. *Izv. Atmosph. and Oceanic Phys.*, **2**(11):1186–1192, translated by A. B. Kaufman.